Electric Power Systems Control Centers

A functional description of supervisory and control systems

Daniel A. Martins

Copyrights © 2018

Daniel Augusto Martins

All rights reserved

To Victor, Julie, Sofia and Luca.

As a stimulus to their fights, challenges, and victories to come.

Thanks.

We would like to thank all friends and colleagues of the "Centrais Eletricas do Norte do Brasil" SA - ELETRONORTE Company who taught me, guided, and assisted me "on the job" for more than thirty years. Thanks also to my conquered friends throughout my academic and professional training. People I hold dearly in my heart. Thanks to all my company fellows, especially to Mr. Sampaio W. V., Almeida C. R., Farias R. do P., Cardoso Jr., J. M., de Pinho D. A., Brito Filho P., Messias A., Corps M.

We thank all professors and especially our mentor teacher Nicole Pouliquen or Nicole-Sylvie Breaud. She made me delve into hybrid computing, the digital electronics arts, and low-level programming that have served so much in my activities as an engineer.

We thank, finally, but also more emphatically to my wife. She has made so much personal effort to enable me to work in an area where the duration of the service depends not only on a company's labor agreement but on the weather, accidents whims of the electrical systems that we are obliged to serve urgently whenever necessary.

Table or Contents

Thanks...5
Introduction..16
    Installation supervising – a bit of history.........................19
    Installation protection system 20
        Circuit breakers and disconnecting switches..................23
    Installation supervision and control....................................24
    Data logging and the remote terminal unit - RTU.............25
Operation of electric power systems..29
    Complex numbers – the complex plane 29
        Number of phases – the number of cables used 30
    The state of the electrical system..31
        Secure or normal state 31
        Alert state 31
        Emergency state..31
        Extreme state...32
        Restorative state 32
    Basic requirements for system operation 33
        We should always try to attain load demands. 34
        Energy quality..35
        Frequency 35
        Why it is necessary to maintain constant voltage?............36
    Regarding the projects purpose 37
        Electric energy generation..37
        Electric energy transmission..38
        Electric energy distribution..39
    The scope of the Operation's responsibility 40
        Local or installation operation...40
        Area operation – the area control center - ACC...............41
        Regional operation – Regional control center RCC...........41
    System operation center..42
        Interconnected system operation 42
    As for the crew specialty..42
    Geographic and architectonic aspects 43
    Centralized Supervision 44
        Control center responsibilities..44
        Operation of the electrical system area under their responsibility 45
        Operation planning 45

> Performance evaluation 45
> Administrative *activities* 45
> Telecommunication system 46
> From minicomputers to LAN 48
> The need for redundancy 49
>> Software redundancy 49
>> Communication redundancy 49
>> Security 50
>> Power supply security 50
>> Installation auxiliary services 50
>> No-break and battery packs 50
>> Programming security 51
>> Robust programming 51
>> Programming standardization 51
>> Invasion security 51

Control Center functional analysis 53
> SCADA – Supervisory Control & Data Acquisition 53
> Analog supervision 53
> What does a control center do then? 59
>> Volatile memory stores variables. 59
>> Variables are offered to be seen by the operators 59
>> Variables are stored in classical SQL databases 59
>> They are transmitted to other control centers 60
>> Proprieties of an analog variable 60
>> Analog variables logic proprieties 62
> Digital supervision 63
>> Events chronology 63
>> Measurement error and security 64
>> Logical variables are stored in memory 64
>> Logical variables are offered to operators 65
>> They are stored in disk SQL databases 65
>> Logical variables are transmitted to other control centers 65
> Command functions for the electric power system 66
>> Logical commands 66
>> Analog commands 66
>> TAP changer commands 67
> The importance of TCP/IP in remote command efficiency .. 68
> Systemic protection 73
> Supervisory services offered by a SCADA 73
>> Static data or metadata and dynamic data 74
>> Data storage 74

Data logging – alarm and event storage 75
Data logging - SOE history 75
Storage of electrical quantities - analog history 76
Analog measurement storage – analog historic 76
Analog history in text files or spreadsheets 76
History stored in SQL databases 77
The amount of data stored - data warehouse. 77
Time synchronization between installations – GPS 78
Single-line diagram – the operator´s eyes 79
Installation topological drawing 80
Analog variables representation 80
Logical variables 81
Audible alarms – the operator's ears 81
Historical graphs 82
SOE on line consultation 83
Automatic voltage control - AVC 83
Energy Management System – EMS 85
Power flow calculation – The Load Flow 85
State estimation 89
State configurator 90
Contingency analysis 91
Load forecasting calculations 94
Economic dispatch calculations 95
Automatic Generation Control, AGC. 96
Operator Power Flow - OPF 97
Summary of EMS functions 97
Control centers interaction to other areas 98
Reporting assistance 99
Service Orders – S.O. preparation assistance 99
Connecting supervisory network to INTRANET and INTERNET 100
Enterprise information sharing 100
Offering services on the INTERNET 101
Integration to a phasor measuring network 101
Operators training 104
Integration to business management system, BMS 104
Operation control Systems and BMS interaction 104
The use of modern mathematical tools 106
Use of artificial intelligence, AI techniques 106
The use of artificial neural networks, ANN 106
The use of outliers detection techniques 107

Stability assessment using decision trees 108
The high number of samples, Big Data 108
Renewable energy 109
   Solar energy 109
   Wind energy 110
   Other sources of energy 112
The smart grid 112
Final considerations 115
BIBLIOGRAPHY 116
About the author 119

Introduction

This book is for electrical engineering students who seek to understand how a control center helps to operate an electric power system. Which tools are used, which are the control center functions, skills, and specializations required by the team of technicians, engineers, administrators, and what goals embarked personnel in an electrical power systems control center in their daily labor? This book is also suitable for electronics and information technology students. Whether out of curiosity or out of necessity, want to know more details of an operation center.

The emphasis of this book is the discussion, from the functional point of view, of the available software in a power system control center. Here we understand the word software as a set of programs, tasks, scripts competing for CPU time to perform their functions. We describe their relations and how this set can assist in the detailed observation of the behavior of the electrical system and assist the operators in performance decision making and reconfiguration commands to increase security, availability, and efficiency in large-scale power supply.

Most of the tools needed to operate an electrical system today are IT tools. They automate and increase the safety of the electrical system. Some tools used are "of the" shelf tools, such as the TCP/IP communication protocol, which are usually already shipped in operating systems. Others are specific to produce computer modeling of the electrical system or perform functions that improve the quality of the electricity supply service.

Living beings cannot perceive electrical phenomena very simply. With hearing sense, some people may perceive the "snoring" of 50 or 60 Hz by, for example, when a sound amplifier does not have its alternating current filters properly working. Closing an electric circuit through a living being can only be perceived when it causes discomfort, physical damage, or death. There is a constant risk in electric power systems where lower voltages, such as 440 volts or 13.8 KV cannot be perceived, not even at close range. Very often they are deadly. The highest voltages, for example, 69 KV end up being less dangerous because, next to them, the hairs of the arms and hair bristle.

The arrival of micro-processed computers allowed the integration of servers through local networks. The rapid growth of information technology -

IT, and the development of digital electronics allowed for the modernization of power plants. For example, the rapid development of power electronics, relay position state sensors, low-intensity currents sensors proportional to high magnitudes of voltage, and current has allowed what we call the "eyes and ears" of electricity. Everything became more comprehensive than the old pointer meters and signaling lights used in the old panels of generating plants and substations. The need for reliable, safe, and uninterrupted operation of power systems has brought work not only to electrical but also to electronic engineers, telecommunication engineers, and computer specialists, both in hardware and in the development of dedicated software. So, this book can be considered multidisciplinary and arouse interest for a wide variety of professionals. To discuss such a large number of subjects in a book forced us to reduce the scope of each theme. Whenever possible or remembered, a bibliographic reference will allow going deeper on some Subjects.

The operation of electric power systems has always been conservative. The Electric Sector in any country in the world is considered as a conservative sector. The risks involved always led cautious managers to doubt the use of non-traditional equipment. It was the 1965 and 2003 blackouts in the United States and Canada that led to the drafting of government recommendations that forced the introduction of computer parks and power management programs in the operation of power systems in the case of 1965 and the definitions of strategies in the case of the 2003 blackout (Wu, 2005).

In 1981 the Brazilian government promoted the formation of the first group of personnel specialized in energy management engineering using sophisticated programs originally developed by PECO - Philadelphia Electric Company that allowed the efficient correction of the measured data, the state estimator, and in the production of answers to "what ifs" questions by the planning tool known as contingency analysis that helps to avoid possible instabilities that are likely to occur if the electrical system changes state more severely. Contingency analysis is a widely used tool in the area of energy planning and will be discussed further in this book in more detail. At that time these tools were not in the control centers. Usually the computers of the administrative processing center were used to develop the modeling and analysis of the behavior of the electric power system.

This book is adequate for undergraduate students in electrical engineering, electrical and electronic technicians involved with specific activities in an electrical power systems operation center. It is also a good choice for those interested in control and processes automation in general. Operation engineers and system operators are also a clientele for this work if they are interested in better technical details about how computers control the electrical power system.

This book describes the point of view of those who purchase the computer park to compose their operation centers, the software to meet the needs of providing power to the population in a safe, uninterrupted, and reliable manner. It's not a manufacturer's vision of electrical equipment off-the-shelf system.

This book describes the experience of more than 30 years in the Brazilian electrical sector. It describes the experience with control center installation, commissioning tests, improvement, and development of supervision and control systems. Control centers allow for the increase in the reliability, safety, and availability of the electric power system through the correct and efficient use of the technological resources available in the operating centers.

To better understand an electric power system, we invite you to appreciate the figure below. It is an interconnected electric power system. Several generating power plants spread all over the country deliver energy to consumers through transmission lines and connecting substations. It is one of the biggest achievements in engineering.

For example, in Brazil, practically all states are interconnected in a single electrical network. We can see the electric power system as the biggest machine in operation in the country. An electrical power system is a large web, like a gigantic spider web made of cables, sometimes through the air gap between the windings of transformers and electric machines. It conducts energy produced by generators situated hundreds, sometimes thousands of kilometers away to our simple washing machine outlet.

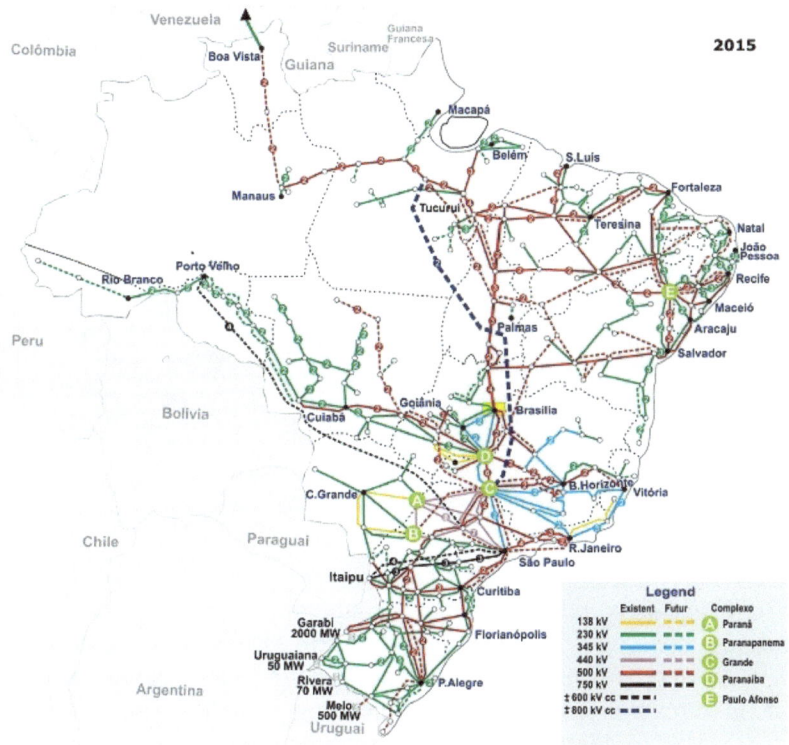

National Interconnected System

How to keep this machine running reliably, safely and uninterrupted using control centers is the result that we aim at in this discussion. (Monticelli a. , 2011).

Installation supervising – a bit of history

So it is necessary to carry the energy generated by an electric generator to multiple consumers. Electric energy distribution started around 1890. Groupings of equipment allowed the redirection of the energy through lines in poles or other support structures. These equipment groupings used to redirect energy are called **substations**.

[Courtyard of an electric power system substation](#)

This grouping of equipment is called an installation or when there is no power generation and only voltage levels transformation and redirection of transmission lines, a substation. A substation is a set of materials. Cables, metal structures, equipment, such as transformers, cable disconnecting equipment (circuit breakers and disconnecting switches), pump drive relays, protective relays to operate in the sectioning equipment, sensors, and bus bars special, called feeds. Electric energy flows from a substation through poles or transmission lines to the various communities served.

Frequently, a substation has various reels, called reactors, sometimes mobile for fine control of voltage. These are called synchronous compensators, and capacitor sets are also used for voltage control, among other equipment.

Installation protection system

In addition, it has long been necessary to add specialized electronic equipment to take the initiative of shutdown when some abnormality was found in the behavior of the electrical magnitudes, thus preventing material and human damages. An unbalance of voltage or current in a three-phase electrical system may indicate the proximity of an accident due to a short circuit in one of the phases.

Raising currents in a line may also indicate an unpredictable and unbearable increase in load that may damage transformer windings with risk of fire, damage to assets, and life-threatening. That is why electrical installations are equipped with electronic instruments, earlier with electro-mechanical instruments. They can quickly trigger the circuit breakers. this is called a protection system.

Old overcurrent protection relay

The figure above illustrates an old protection instrument, an over-current old protection relay.

A whole knowledge area was born, called Electrical System Protection. It teaches the best strategy of installation and adjustment of protection equipment.

Even in our houses, a rudimentary form of protective equipment is used. They are the building fuses and breakers in the power boards.

The figure below illustrates a much more modern, microprocessor-based and fully software-configurable protection equipment that combines several functions of protection, supervision, control, data transmission and time synchronization via GPS - Global Positioning System.

The protection strategy has been sophisticated and its efficiency has increased over the years. Concepts such as the one of teleprotection in which a signal is sent from a substation, that perceives the risk, to a neighboring substation, requesting that it acts in the opening of a circuit breaker has been naming these strategies as "philosophy of protection". Electric protection is a vast study area that, until the time of these writings, has not been given importance in electrical engineering courses. For over 30 years it has rarely been offered as a standard discipline in undergraduate courses. In general it is optional both in undergraduate and graduate studies. Most of the experts in this area that we contacted in our years in the Brazilian Electrical Sector were formed in the company itself trying to solve and improve protection systems delivered by suppliers of equipment in substations and generating plants. Some have made it as an elective discipline.

Circuit breakers and disconnecting switches

It is worth noting the difference between circuit breakers and disconnecting switches. The routing of the electric current in a substation to the various consumer areas or the following substations, selection of the generation source or the previous substation supplying power to a substation is performed through circuit breakers and disconnecting switches. The circuit breakers are very fast drive switches, usually a few hundredths or thousandths of a second. The circuit breakers are thus triggered by the protective equipment to quickly disconnect an area of the substation, thus avoiding material or human damages. The protective relays are then the instruments that "make the decision" to open circuit breakers. The closure of them is usually by human action. Nowadays they are driven, also, by more modern automatisms. The switch, on the other hand, is constructed with large metal rods, often with meters in length to allow to electrically isolation of a substation area by distancing the energized point. Thus, the isolating switches are insulation equipment. Its activation for opening and closing can be by command but they are equipment used by maintenance personnel. Whenever there is a need for human intervention in an area of the substation yard, the nearby switches are open to allow staff to be present without risk.

A three phase circuit breaker to the left and disconnecting switches to the right

The figure above shows an example of a circuit breaker and a switchgear in a high voltage substation.

Installation supervision and control

The complexity of the installation, the exposure to weather, and the voltage levels present already suggest a great difficulty to operate it directly from the yard outdoors. The life threat involved when walking under high or extra-high voltage cables suggests that the conduction of sectioning equipment and observation of the electrical values that circulate or are present in power cables and bus bars should be effectuated at a safe distance. Hence the use of command houses and the necessary wiring interconnecting the command houses to the yard. Historically then a commanding house or a control room was composed until recently by a group of metal cabinets containing three types of panels: a panel of lamps representing the state of sectioning equipment and protection relay states; a panel with analog pointer meters showing the values of voltages, currents, energy flows, etc. and finally a set of buttons and keys that allow remote activation of circuit breakers and switches from the control room, thus avoiding dangerous frequent visits to the patio. A control room for an installation can already be considered a control center.

[A facility control pannel](#)

The figure above illustrates found in old control panels to supervise and command a bay inside an installation. Below the remote control panel using the pushbuttons for the tripping of the circuit breakers, at the center the panel of alarms consisting of a set of lamps representing the state of the equipment and alarms and protection relay states, and finally on the high the analog pointer meters. It was the man-machine interface available for decades to control the electric power system.

Data logging and the remote terminal unit - RTU

The first additional requirement besides the primary visualization of equipment states values and command buttons made possible by technological advances implemented in installations was to register changes in equipment states and include the moment changing occurred. More important than changing time is the chronological order in which the various alterations, openings, closings, actuations, shutdowns, protection equipment actuation, etcetera occurred. The exact changing time is not relevant. The date of the occurrence is not a must. The occurrence chronology, the temporal order in which they happen, is of fundamental importance for efficient global analysis of an installation occurrence, its causes, its whys, and what to do to prevent them from happening again. In this area of electrical engineering, called the analysis of the performance of the protection of electrical power systems, the word protection is also involved in human protection. That is the main objective of this science and art.

In the early 1960s, minicomputer cost reduction and availability of impact printers provided technology that addressed the chronology of events and alarms occurring at a localized substation.

Cables connecting yard equipment to alarm panels inside the command house connect now to minicomputer digital inputs. A program in this minicomputer associates these contact changes of state in the yard to computer variables. These changes immediately logged in a human-readable sequence of lines describing the history as sensed by the minicomputer. This list, locally logged and then sent to protection experts. This implementation marked the beginning of information technology - IT in the operation of electric power systems. The figure below illustrates a remote terminal unit - RTU and its data logging printer. This list is known as the "sequence of events" or SOE - sequential of events.

Remote terminal Unit – RTU and it´s impact printer

Recently PLCs - programmable logic controller - substituted RTUs. A simpler programming language replaced the old RTUs Assembler. This programming language is called "Ladder Logic". Today PLCs accumulate the functions of a classic RTU with command and protection functions. PLCs have the advantage of being modulated. Plant expansions and changes are now easily absorbed without major technological changes. Extending software using ladder logic is also simpler. These changes, when performed in the RTUs were usually sources of headaches for the IT staff responsible for their maintenance. Expanding an Assembler language program is no easy task. The figure below illustrates a typical PLC.

[A programmable logic controller](#)

A standard procedure in an installation was to send this list to experts who were not frequently in command houses after, for example, a shutdown occurred. So, this procedure delayed protection analysis. This list is universally known as SOE - sequence of events. It was, for several decades, the only tool used to analyze occurrences in electrical power systems. It is worth mentioning here that sheets of paper stored the installation history.

These loggings were stored for eventual future analysis. They are a comparison analysis between present to past occurrences searching to understand their causes and how to avoid them in the future. All this paperwork and eventually lots of precious information are now completely lost. For a long time, SOEs were also stored on magnetic media.

UTRs, PLCs, impact printers, and even ink pen recorders began the era of local operation centers in the substations.

Operation of electric power systems

The following are concepts and definitions used by electrical power systems specialists. Some of them differ slightly from country to country, from school to school, but the basic idea holds. A detailed discussion of areas involved with each of these concepts is out of the scope of this book. To better understand what an operation center controls, we think it is necessary to know at least the jargon of the electricians and some definitions, electrical characteristics, and formulas considered fundamental to the observed and controlled system.

Complex numbers – the complex plane

Electrical power system quantities, such as voltage, current, power, etc., are represented by complex numbers. A complex number is a vector in a two-dimensional coordinate system, also known as the complex plane, as shown in the figure above. If we imagine the complex number $Z = a.\cos(\Theta) + j.b.\sin(\Theta)$, where j the angle Θ that this vector makes with the real axis is, as can be seen in the figure, $\Theta = \arctan(b/a)$. Some properties of electrical quantities are:

- If we see complex numbers as a math function we note that it is a linear function. That is, $f(\alpha.x) = \alpha.f(x)$, $f(x + z) = f(x) + f(z)$...
- The complex number resulting from the division of two other complex numbers is the product of the numerator and denominator by the conjugate of the denominator. It is worth verifying by an example. Kirchhoff teaches us that.
- The product of a complex number by its conjugate is equal to the sum of the squares of each term, real and imaginary.

So, if $Z = a + j.b$, then $(a + j.b) \times (a - j.b) = a^2 + b^2$

- The conjugate of a complex number is a linear function.

In electrical power systems, the complex power **S**, or apparent power, is measured in millions of ampere-volt or megavolt, MVA, or in KVA for distribution systems.

Like every complex number, it has a real part and an imaginary part. The real part of apparent power is called active power and is measured in millions or thousands of watts, megawatts, or kilowatts, MW, KW. The imaginary part, or apparent power, is called reactive power and is measured in millions of reactive ampere volts or MVAR or thousands of reactive ampere volts KVAR. These two quantities play a prominent role in electrical power systems.

Electricians know that MW concerns useful electricity consumption. MVAR is strongly linked to the voltage at substation bars, and therefore, a strong indicator of stability problems, as will be discussed further below.

MVAR, or KVAR, exists because the voltage produced by a coil generator does not have de same phase angle as the current. Photovoltaic generation, for example, voltage and currents generated are continuous and therefore have the same phase angle, zero.

For the supplier company, reactive power means loss of transmission, and typically it charges the consumer because it is his responsibility to inject reactive in the grid. MVAR or KVAR is caused by the use of motors in industries or at home. For the consumer, reactive energy injection is minimized using compensating capacitors to reduce their monthly bill value.

Number of phases – the number of cables used

Electric power is transmitted from its generation to the final consumer through three conductors called phases. An additional conductor is called neutral or ground. It is usual to name each conductor with letters A, B, and C or still as A, B, and V. The wave shapes are sinusoidal. This mode of transmission is called alternating current transmission. Only when transmitting large amounts of energy, when the voltage in the conductor approaches 1,000,000 volts energy is transmitted using direct current. The detailed study and mathematical modeling of AC and DC electric energy transmission is a very specific area of electrical engineering knowledge dominated by very few.

The state of the electrical system

The set of voltage values, phase angles of these voltages; line power flows at several points of the electric power system is known as the "state" of the electrical power system. This set may assume normal, slightly abnormal, or alarming values. Specialists eventually divide the behavior of the electric system into three states:

Secure or normal state

All variables are within reasonable ranges. The system runs at no risk of collapse. For example, voltages should vary by ± 5 % of nominal values to be considered normal. Flows should be around nominal values recommended by equipment manufacturers. Voltage and current values measured at every one of the three phases should be close to each other.

Alert state

Some variables may leave their normal limits. For example, an electrical disturbance in the electrical system result, the power system may go to an alert state. Generator or consumer disconnections, high load variations of some consumers, power line disconnections can lead the electrical system to an alert state.

Emergency state

A severe disturbance such as the loss of several generators, disconnection of a connection point to another area, and severe loss of charge can lead the system to an emergency state.

Extreme state

Cascade shutdowns, blackouts in various areas can lead the system to collapse. Whole power plants shut down. Transmission lines disconnect by protective equipment, and thousands of consumers run out of power.

Restorative state

Operators act to reconnect the generators and synchronize them with the network. Operators reconnect the transmission lines to restore loads.

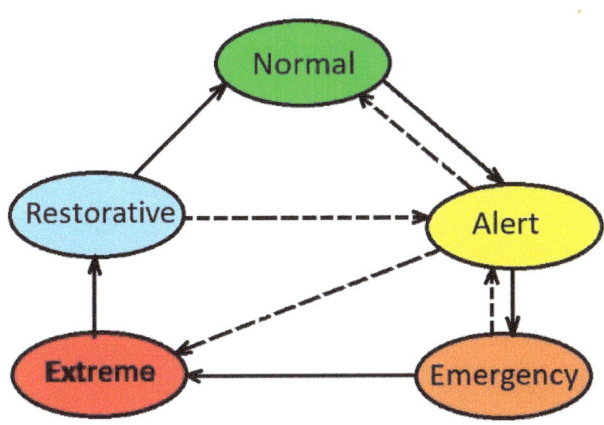

The states of an electric power system

The figure above illustrates the paths taken by system operators when a disturbance occurs. In the normal state: relax. Operators have nothing to do except planning studies and eventual operational, administrative reports, and constant surveillance. The supervisory system remains alert. We expect the system to be 100 % of the time in a normal state. In the alert state, the stress level starts to rise. We then make arrangements to adjust voltage levels.

Possibly some switching maneuvers for connecting reactors, capacitor banks, and fine adjustments in synchronous compensators and other adjustment equipment are taken into effect. In emergency state stress level gets even higher. It can mean loss of consumers and losses of generators.

Operators take more severe measures or try to reconfiguration the system. Operators are either in the high-stress state or sometimes in a near-panic state, trying to get the system back to normal. Sometimes we take even more severe measures, such as load relief, which may lead to the withdrawal of elective consumers as non-essential, triggering or shutting down generators or power plants, etc.

In extreme states, rare to happen, expected only in blackouts the fight is to minimize disaster Try to adjust the system at all costs to take it back to the restorative state where the system begins to recover gradually and safely. Of course, these actions should prioritize human security before asset security.

Blackouts, which characterize the extreme state, are the price we eventually pay for having the system interconnected. The interconnection of the system has brought financial benefits and reduced investment needs to meet the growing and natural demand for energy in the country. Brazil cannot fail to use this policy because it needs to count on the blessing of owning watersheds with complimentary seasonal behavior.

When rivers in the south are voluminous, the rivers in the north are diminished and vice versa. There are times north region uses energy from the south; other times south uses energy from the north. Canada is a country more latitudinal than longitudinal. It cannot use so much of this geographic gift. The price we pay is that the more interconnected, the greater the possibility of contingencies spreads between regions.

The interconnection of the system favors the risk of instability and points in the direction of the blackout. Stability analysis of electrical power systems is a significant part of electrical engineering knowledge.

Basic requirements for system operation

We then discuss the more general goals pursued by operators in their control centers. General requirements are here summarized.

We should always try to attain load demands.

Domestic and industrial consumers suffer a seasonal energy variation that depends on the hour of the day, the weekday, and the year season. We show below a typical curve of daily energy demand.

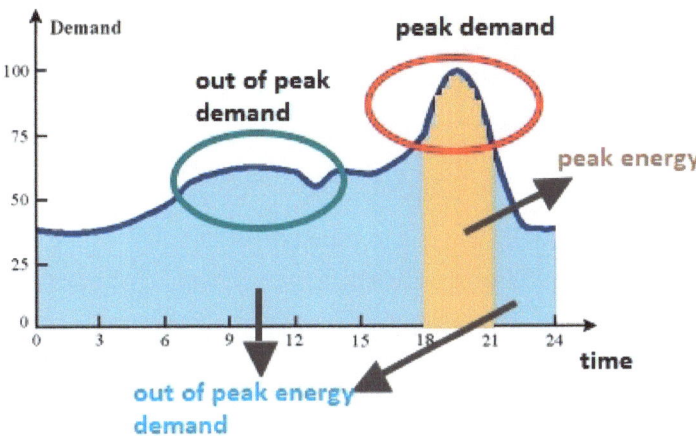

Tipical daily energy demand

The electrical system must meet with clearance the peak demand, which usually takes place between 18 and 21 hours depending on the country, season, and daylight saving time. If it does so, it works off duty the rest of the day. The term "clearance" may seem an advantage at first sight. However, it indicates that investment is required to meet demand peak value. Nevertheless, the investment would be lower if the curve was flat. The asset would be lower if it were possible to reduce the peak value.

The dream of all energy companies is to reduce consumption at peak times. We will discuss this subject further ahead. The economic growth of the country with new industries, and population increase, etcetera leads to an average demand year-by-year increase. Thus, constant investments are needed to meet growing demand.

Energy quality

There is a growing concern with energy quality nowadays. A large number of components and equipment with non-linear behavior such as the most modern lamps the electric brush motors etc. inject a great number of harmonic components in the electrical system. The system is designed so that its electrical waves are pure sinusoidal waves of frequency 50 or 60 Hz. Components of frequency of 120, 240, 480 Hz, etc. are called harmonic components. These components can, for example, penetrate the power supplies of precision equipment, such as computers, tomographs, radars, etc., despite the presence of customary filter capacitors.

A recent area of study for the analysis of harmonic components has been growing and producing results. Instruments, called "qualimeters" are available in critical areas and are interconnected to control centers to offer and store the amplitude of harmonic components in order to allow their analysis, to determine origins, causes, and how to reduce or avoid them. Harmonics are the undesirable noises of the electrical system.

Frequency

Why maintain constant frequency? Frequency fluctuations are harmful to home electrical equipment. The electric sinusoidal wave frequency produced by motors in alternating current is proportional to the rotation motor speed.

$f = (N \times p)/120$ where,

f = electrical grid frequency

p = number of generator poles

N = motor rotation speed in RMP

Water flowing through a turbine is not constant. There is a speed variation, and therefore, a network frequency variation. Electromechanical controllers correct these speed variations.

In addition, the difference between the measured value of the frequency with respect to the base value, 50 or 60 Hz is an indication of imbalance between generation and consumption load. Thus the frequency imbalance is a numerical representation approximately proportional to the energy imbalance or to the equilibrium of the system or to the stability of the system. Frequency above the nominal value indicates generation greater than demand and frequency less than the nominal value indicates that generation is not being able to meet demand.

Since the 1950s, when electrical systems began to expand, a sample of the frequency and power generated by the various generators scattered across a region were taken and transmitted by analog radio equipment to combine to consumers load. It was the beginning of the charge-frequency control. An attempt to perform what is called today's optimal dispatch charge. This arrangement, we can consider as the oldest function of an electrical power system control center.

Why it is necessary to maintain constant voltage?

Overvoltage

The difference between a measured frequency and the 50 or 60 Hz base value represents the imbalance between generation and load.

Thus, frequency unbalance is approximately proportional to the system generation-load system. It is a measure of system stability. Frequency above nominal value indicates generation greater than demand, and frequency below nominal indicates that generation does not meet demand.

Electric motors tend to spin at over speed when powered by voltages above normal. So, Over speed results in vibration and increases the possibility of mechanical damage and aging.

Under voltage

For a given required power, under voltages result in reduced motor winding currents which causes vibration problems, reduced rotational speed, aging etc.

P = V.I, where,

P = electric power

V = coil applied voltage

I = coil current flowing

Supply availability

Consumers pay for 24 hours day energy. Regulatory agencies establish Standards and recommendations. Sanctions are applied when they fail to comply. Consumers and assets must always have comfort and safety.

Regarding the projects purpose

Electric energy generation

Electric power must be generated, transmitted to the large consumer centers, and distributed to customers. Usually, the available site to generation is far from great human centers. Rivers with high slopes or waterfalls are the best places to build Hydroelectric power plants. We construct wind farms onshore or offshore. We use sunny-wide regions to build photovoltaic parks.

Electric power generation is often very far away from consumers. We overcome this distance using copper or aluminum cables supported by metallic towers in high voltages to minimize losses by heating caused by the electric current. These towers are high enough to isolate the cables to the ground. We avoid short circuits caused by high voltages. Several substations are often necessary to overcome distances as the voltage drops as current flows through the cables. Substations, then, serve to redirect energy to various consumers. Substations also restore voltage levels to adequate nominal values.

The voltage level is low in consumption centers. In Brazil, it is reduced to 69 KV and distributed in even lower values, for example, at 13,800 V, 440 V, and down to 110 or 220 Volts for residential consumers. Traditionally, regional concessionary companies are responsible for consumer energy delivery. We can divide energy companies by their competence of acting as generation companies, transmission companies, and distribution companies.

A common type of generation in Brazil is hydroelectric power plants which convert the potential energy obtained by the height of the water column obtained by the damming of a river into electric energy through the flow of water by blades or propellers that drive the coupled turbines to electrical generators. Thermal power plants convert heat from oil combustion, coal, nuclear reactors, etc. We use hot water at very high pressure to move turbines which turn the electric generators producing energy. This form of generation is called thermal generation. Nuclear plants are also considered thermal plants. That white smoke that rises from nuclear generators used in the press to illustrate environmental pollution is just water vapor. The danger is below.

New ways of converting energy into electricity are increasing nowadays. Photovoltaic plants convert the energy from the Sun to electrical energy. Wind power plants use wind force. Tidal power plants use the sea wave or tides force to generate electric energy.

These new ecologically adequate forms tend to replace thermal or hydraulic generation to diminish environmental impact. Farms, industries, and even homes are using photovoltaic energy conversion. Germany is the best example of how to get more independence from distribution companies and during some hours of the day and sell surplus energy.

Usually, in hydroelectric and thermal plants, power generation is produced at a voltage level in the vicinity of 13,800 volts. Transformers raise this voltage in the substation, adjacent to the generating station for transmission.

Electric energy transmission

Transmission companies' responsibility is to transmit energy from the generation source through transmission lines and substation's consumption centers. They are companies specialized in Electric Power Transmission. Efficiency, high availability, and safety in energy transport are the dominant concerns of these companies.

Power transmission is performed at voltage levels generally ranging from 34.5, 138, 230, and 500 KV. Nowadays, energy transmission beyond 700 KV uses direct current. The classification of voltage levels varies slightly from country to country.

Electric energy distribution

It is the responsibility of the distribution companies to promote the delivery of energy to the final residential consumer and eventually to special ones such as industries and large companies. These are companies are concerned with the transmission of the already transformed energy to lower voltage and using support in poles or underground ducts to deliver it to the final consumer. Here again, the amount of necessary cables rises exponentially since each consumer needs particular wiring. The first concern of distribution companies is the quality of the delivered energy, such as adequate voltage level, energy availability, and the minimum number of interruptions to each consumer.

For the three areas described above, government regulatory agencies establish performance figures to be pursued by companies. For example, it is common to find figures such as:

Equivalent duration of disconnection per consumer

DEC, in Brazil, is the duration a given consumer has no electric energy in a month. A contractual maximum time interval value is established, and sanctions are applied.

Equivalent frequency of disconnection per consumer

FEC, in Brazil, is the number of times the consumer was subject to the interruption of electricity. FEC is also accounted for a period, usually one month. When the concessionary does not attain a maximum month number of losses, sanctions are applied.

Regulatory agencies set different values for DEC and FEC for either, transmission and distribution companies. This pair can also be found with the names DEK and FEK, depending on the company. Energy companies (CEMIG, 2009)Balance sheets establish DEC values of 77.5% to 98.5%.

Normally power distribution is carried out using voltages below 69,000 volts, inclusive. Voltages of 69,000 volts, 13,800, 440 volts, 220 and 127 volts are consumer used voltage levels. The latter normally reach the entrances of companies or residences, the final consumers. Industries, major consumers, can receive up to 13,800 volts.

Attention to non-electricians! We hear people talk of voltage values of 380 volts on street poles for three-phase transformers, therefore, with three windings. Beware there is a square root of three hidden in this statement. 380 is the voltage value between phases. The phase-to-neutral voltage is achieved by $380 / \sqrt{3} = 220$ volts. And even more, 220 volts, the voltage offered to us at home is the phase-phase voltage or the difference between phases. At our outlet, we may also receive $220 / \sqrt{3} = 127$ when using voltage between phase and neutral in many countries.

The scope of the Operation's responsibility

For the operation of electrical systems, we establish various levels of responsibility. Personnel of numerous specialties attend to the complexity of substations and power plants from dispatch centers, coordinating and planning power flow between the different areas or companies. Nowadays, with technological advancement, discussed in more detail below, some facilities are left unoccupied but remotely assisted. One or two operators are usually present during office hours. In some circumstances, they are kept nearby, ready to intervene in unexpected occurrences. Typically control centers range is classified in:

Local or installation operation

Operation is local when it is responsible for a plant or substation. Operators follow written guidelines to operate safely, reliably, and continuously the installation under their responsibility. We mean by continuously non-disruption. Increasingly nowadays, the local operation centers are left without having a permanent crew.

Voltage regulation and programmed switching are carried out by computers nowadays. It is common to find out a specialist only in the vicinity of small substations or hydroelectric plants. Wind towers, photovoltaic parks, for example, are already born without a crew but not unassisted. Assistance is performed automatically remotely or locally in the case of communication loss.

A particular local operation center is the plant operation center - COU. This center, in addition to the characteristics of the other operating centers, uses specific functions for, for example, treatment of remotely received basic power consignments to be supported by the generator group of the plant and the equitable and strategic generation distribution among them.

Area operation – the area control center - ACC

Area operation centers are responsible for an installation set, spread by a geographical area, planning, and operation. They are responsible for coordinating a group of installations. They are also engaged in post-operation analysis. We establish an energetic dispatch planning to attend this installations group. Planning takes into account the geographical characteristics and regional state policies.

Regional operation – Regional control center RCC

Tasks and concerns are now related to the coordination, operational planning, and post-operation analysis of a group of power plants and substations in a group of states associated with a particular geographic or political region. Now, electro-energetic planning takes into account a whole country region. It takes into account geographical and political regional characteristics. These operation centers are now called dispatch centers. The word dispatch designates planning, monitoring, and altering the amount of energy transported from the origin to the final destination.

System operation center

Responsibilities are now related to energy planning, dispatching, and load coordination between regions and to other companies. Planning operation in addition to the pre-operation planning and post-operation analysis of all operation centers now take place. In these operations, centers load dispatch activities are carried out by specialized areas and involve a much larger number of experts in planning, operation, protection, and post-dispatch analysis. It is here that occur exchanges of information with the various other areas of the company. From the area of law to economic, to accounting, to economic planning, to relationships with other companies now take place.

Interconnected system operation

The responsibility is now national. Global energy planning, global dispatch and maintenance of energy levels and flows are a taken place. In Brazil, National Operator - ONS carries out the nation-wide operation. It defines network procedures. It determines a set of operational standards and technical requirements necessary for the country's safe electro-energetic operation. ONS is also the responsible real-time operation of the whole country. It conducts what we call National Integrated Electric System - SIN.

As for the crew specialty

The operation of the electric power system is a multidisciplinary activity. It requires experts from various fields. The result of the crew effort is what ensures efficiency in the energy supply to final consumers. In energy companies, there is usually, personnel with the following specialties:

- Electrical and electronic engineers and technicians
- Engineers and technicians specializing in electrical protection
- Engineers and technical experts in the operation of the system
- Engineers and technical experts in telecommunications
  o Economists
  o Lawyers
  o Administrators, etc.

Geographic and architectonic aspects

Regional or local control centers are often found near power supply. Local or regional control centers are found near substation courtyards, at hydroelectric plants or near wind farms. Regional Control Centers - RCC are most often located in state capitals and System Control Centers - SCC in the nation's capital.

The building's architecture usually includes a large room, the operating room, where the electrical system operators are located in front of their multi-monitor terminals used to view single-line diagrams, alarm lists, and administrative work. Usually the operating room has a large panel, the synoptic board, formerly made with drawn lines and bright beacons and today assembled from large video monitors to form even larger images.

Buildings typically also house rooms adjacent to the operating room where the nearest technical personnel are located, such as a room for electrical planning, pre-operation personnel, and incident analysis and system protection personnel. We find a post-operation room, and possibly also rooms for the crew that takes care of the computer system and the engineers and telecommunications technicians.

The photo below is an example of an operating center's control room of Eletronorte Regional Operating Center in Belem.

Regional Centers – ROC typically have two operating tables for shift operators and one operator coordinating table. In system operation centers - CCS, the number of tables may be larger, for example, tables for operation engineers using the power management functions - EMS.

[Eletronorte Regional Belem Control Center.](#)

Centralized Supervision

All equipment needed to supply energy spreads across vast regions to attend the complexity of the electrical system. The arrival of technological advances in power electronics, digital electronics, and industrial computing led to the integration of computerized equipment acting as an aid in the operation of the systems. Advances in telecommunications, especially the arrival of optical fibers, which found a natural path inside the ground cables at the top of the transmission towers, allowed the secure integration of a computer park, usually interconnected by efficient and reliable communication protocols such as TCP/IP. All control centers exchange information with each other.

Control center responsibilities

As already discussed operation centers are located in the facilities or substations, in the coordination of regions, in the headquarters of the companies have their main responsibilities described below. A detailed description of the past, present and future of a control room in a power system operating center can be found in Wu.

Operation of the electrical system area under their responsibility

These are activities carried out by operators in work shift. As already discussed, some installations are unmanned but remotely assisted thanks to the use of computer and telecommunications equipment. These activities are carried out in real time. The coordination of maintenance activities in the field is the most important operation activity. The general "visibility" of the facilities provided by local control centers makes it possible to keep yard maintenance personnel minimally exposed to electrical and physical hazards.

Operation planning

Usually referred to as pre-dispatch, when operating shifts are scheduled, maintenance activities and programming received from higher-level energy planning. These activities are performed by specialists in operation, but no longer in shifts, using the normal working hours.

Performance evaluation

Performance of the electrical system is under its responsibility at the end of periods. This is usually called post-dispatch. Here the supervisory system data is used mainly to elaborate periodic performance reports. Great effort is concentrated by the personnel specializing in the protection of the electrical system to prepare reports of advice on possible changes in procedures and adjustments that prevent the performance of the protection when occurrence of disconnections happens during the period.

Administrative *activities*

Necessary ordinary administrative activities of the company carried on. Included here are recycling activities, training, relations to customers, accounting, financial activities, etc.

The whole effort of the areas described above, we must emphasize, is to maintain the reliability, quality, and availability of the electric system assets and human security.

Telecommunication system

A Local Area Network, LAN, is responsible for the interconnection between the operation centers. The available telecommunications systems interconnect centers. Communication radios were used, historically, for long-distance transmissions. However, they did not use the air as the transmission media. They profited from the long-distance lines available interconnecting installations. This method of communication used a "carrier wave" the Carrier. The figure below shows a typical Carrier coil at the input of a substation.

A carrier Coil

The signal is modulated at a frequency much higher than the 50 or 60 Hz energy transport frequency in this communication system. This signal is transmitted and received by an electronic high-pass filter at the inputs and outputs of the installations. This system is known as carrier or carrier wave transmission. The high-frequency filter installed on each side of the transmission line at the inlet or outlet of an installation is called the carrier filter or carrier coil. The figure above illustrates an example of the carrier coil at the input of a substation.

We transmit data by the ground cables within which we find optic fibers. It is the highest cable interconnecting two towers.

These cables are called OPGW "optical ground wire" cables. They are responsible for all communication, voice, video, intranet transmission. Today energy companies act as telecommunication companies as they sell optic fiber traffic. The figure below illustrates the positioning of the layers of an OPGW cable.

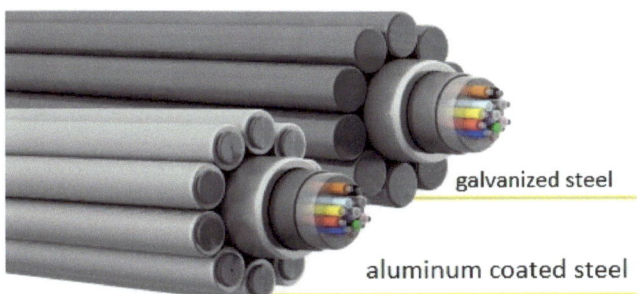

An OPGW1 cable

Currently, almost all companies use the TCP/IP protocol for data transmission. Administrative information adds to the electrical and commands control and supervision information. It is common to find other specific protocols embedded in layers of TCP / IP, mainly in the application layer as voice-IP, video, etc.

From minicomputers to LAN

Historically, the regional and systems operating centers were equipped with minicomputers, usually in pairs, to increase the availability of functions through redundancy. A heap of programs, sometimes written in Assembly language, were offered by the companies that sold operating centers to implement supervision and control of the electrical system on those computers.

There were no commercial supervisory and systems later on. Each manufacturer offered several features to meet the requirements of safe operation. Software development was "à la carte," implemented from a document usually called a "functional analysis" prepared by the contracted company following a "basic project" issued by the contracting company according to its operational needs and meeting a project of the energy company. The characteristics, software development pattern, and deployment are negotiated from approval to program development, testing, and operation delivery.

Developed programs need to communicate with each other to meet the required functionality. They must also communicate with other computer programs and specific equipment such as GPS for time synchronization, graphic recorders, redundancy computer, etc. The set of programs, the development environment, the operating system, the compilers used, and the software libraries must meet computer concepts such as real-time programming, time-sharing, etc., to achieve harmony. Up to the present little is said about the use of object-orientation, encapsulation, and heredity in programs that perform heavy work in a control center. C, ANSI C languages are used in programs and tasks that make up the control center software core. Even C++ is not much found.

Today, operating centers are equipped with more than two computers to assist the operation of electric power systems. LAN networks - "local area network" - are usually adopted. Programs are duplicated or multiplied by several network nodes to increase redundancy. Strategies for the intelligent distribution of values of electrical quantities among the various nodes are adopted to ensure the integrity of the information.

We cannot tolerate a given electric variable with different value in different nodes, even if by a few thousandths of a second.

These LANs usually interconnected the various centers using TCP/IP. The result is a Wide Area Network (WAN) using the available telecommunication means such as radio, broadband, optic fiber, satellites, etc.

The need for redundancy

A tool used to control a system as complex as the electrical system where patrimonial and human risks are involved must be a reliable and always available tool that supports possible computer hardware or software damage. Ready to assume reserve control centers are now in place in remote locations to fully substitute functions of another in the case of natural disasters or terrorist actions.

Software redundancy

To defend themselves, control centers duplicate or multiply critical supervisory functions among the various nodes of the network. It is not an easy task to achieve. For example, a critical process executes in one node, and its twin process sleeps in another node. If the active process dies, it is to be replaced by its twin, and it must use the same data of the death with no loss. Fast tasks communication between different nodes is required. We use the word task to denote a program that runs in real-time.

Communication redundancy

It is common to find redundancy between the computers interconnections to increase system availability on the local LAN or the WAN area network. This last one, due to the high investment, does not have so frequent use. There are several OPGW cables are found in an electrical system with several substations are interconnected. Redundancy, in case, can be more easily implemented. Alternate communication paths leverage the ability to route TCP/IP information.

Security

Control centers are always protected by reasonable patrimonial security schemes, avoiding vandalism and, more recently, terrorism because of their social and strategic importance. Control centers, power plants, and substations are part of the concerns of a country with national security. Controlled access by biometrics or passwords, cameras scattered throughout the buildings are examples of security schemes found in control centers.

Power supply security

An installation of the size of a generating plant or a power substation cannot depend only on one power supply. It is necessary to maneuver circuit breakers and close switches to recompose the substation. In the absence of a power supply of the transmission lines that feed a substation, Generators powered by hydraulic turbines need DC power for the excitation windings to start the generators for electric power generation. Therefore, both a power plant and a substation necessarily need an alternative power supply. Ancillary systems provide this service. They can be found in every installation and provide mostly a secure 125 VCC power supply.

Installation auxiliary services

Auxiliary services are composed of a non-interruptible set of one or two diesel motors coupled to their generators and a battery pack controlled by an "intelligent" switching system. This set provides a 125-volt direct current to feed the remote control panels in the control rooms in the substations and power plants. It provides the excitation current to the power generators in the case of power plants. Control center computers share these asset benefits. In this way, the local control center computer system benefits from the presence of auxiliary services.

No-break and battery packs

We power the control center with the auxiliary services if there is a geographical substation or near a power plant. In the absence of this, it is necessary to use one or two robust sets of UPS – uninterrupted power supply to not allow functions of a control center from shutting down. This uninterruptible power supply system must cover the telecommunications system. The strategy of using both a UPS and an auxiliary service is frequent in control centers.

Programming security

Software developed to perform functions in an operation center must be high-quality software. Internationally known companies specializing in programming for control and supervision of systems are responsible for software development. We must remember that thermal or hydraulic energy high-voltage equipment already poses a high risk for human safety. Risks are even higher at nuclear power plants. Unsecured software can lead to catastrophes.

Robust programming

Programs or tasks developed to an operation center must be built with strict safety and standardization criteria. A high number of comments in source code facilitate their understanding by several programmers eventually hired for software maintenance and expansion.

Programming standardization

We must avoid bad programming habits. Using short name variables, very common to find, cannot be used. Lack of proper indentation in source programs makes it difficult to read. It is always necessary to check program clarity and robustness in the tasks that attend an operation center. We suggest using large software development norms and bring them to our programming habits. (Microsoft, 2016).

Invasion security

To prevent computers in control centers cyber-invasion we must provide proper security measures. We remind you that the safety of control centers is part of national security. In Ukraine, a cyber-attack left ¼ million people without power two days before Christmas 2015. (Greeberg, 2017).

A classic book in Brazil that brings an excellent discussion on electrical systems is (Monticelli a. , 2011). Here we essential discussion of electrical power systems. This introduction serves as a basis because it describes the object and purpose for the existence of a control center. From here on, we will have a less electric connotation and more computing on the subject matter.

Control Center functional analysis

From this point on, we begin to describe the classic functionality of a control center from the computer software point of view. The Control center is the central nervous system of an electric power system. It feels the pulse, adjusts his conditions, coordinates his movements, defend it against exogenous events. Control centers specifications in North America follow a government recommendation to increase computer operating security in power grid control after the 1965 blackout. Brazil followed this example in the early 1970s. (Ankaliki, 2011).

SCADA – Supervisory Control & Data Acquisition

SCADA from Supervisory Control and Data Acquisition is the term universally used to summarize the basic functions of an electric power systems control center. More precisely, SCADA is the name given to the set of programs and functions executed in an operation center.

A happy phrase from a colleague from Brazilian Electric Sector defines well a SCADA system. A SCADA system is the electrical system operator's eyes and ears. We ask permission to extend the affirmative and add the hands. It allows you to feel system changes and also offers the possibility to control them. Supervision is accomplished by obtaining two types of quantities described below.

Analog supervision

Hundreds of thousands of analog sensors are often in action at one facility. Digital to analog converters produce integer binary numbers proportional to the values of the electric quantities in the courtyards from the sampling of voltage values or current values obtained from lowering potential transformers (TP) or current transformers (TC) and other sensors such as frequency, temperature, vibration, level of water, etc.

Sampling period

These samplings occur periodically under the command of the RTUs software, PLCs, and recently microprocessor-based multifunction relays. The time interval between sampling varies according to the type of analog quantity. The sampling frequency may vary from a few minutes, such as for temperature sensors up to a few seconds, for voltage and current measurements up to one to two seconds for network frequency measurement. Sampling could also occur by detecting the variation of each magnitude. In this case, it would cease to be periodic. At the limit, sampling could occur whenever at least one bit of the A/D converter changes.

Calculated variables

In a control center, some variables are not obtained directly from sensors in the field. We calculate them from other variables. It is the case, for example, of active powers in MW, reactive in MVAR, and the energies active in MVAh and reactive in MVARh that can, for example, can be derived from the voltages and currents.

Sometimes these integral binary numbers proportional to the values of the quantities are already converted by the computer equipment responsible for obtaining them in the field. Constantly, the control center receives raw values. The first service of a control center is to convert these numbers into actual values (floating point) for use and storage. The conversion function is, in most cases it is a linear conversion.

[Real value] = [binary value] x [conversion factor] + [adjust constant]

As the equation above shows the conversion formulas are linear overwhelmingly a straight equation.

Periodic or random sampling

It is interesting to note that, although the sampling in the field is periodic, to the computer, it behaves as if it were random since it is not the control center that dictates sampling instants. Thus a control center must have computational firepower to support thousands of binary numbers randomly arriving, representing magnitude values in the field. We use the word field to designate plants and substations, everything that comes from the yards of the facilities.

Measurement error

The figure below illustrates the sampling process of each analog variable in the field. For example, a line voltage at 69,000 volts is lowered to a few volts by the input TC transformer. This voltage is then proportional to the high voltage. It goes through a low-pass filter represented by the RC circuit. The resulting filtered value passes through an analog to digital converter that converts this voltage level to a binary integer.

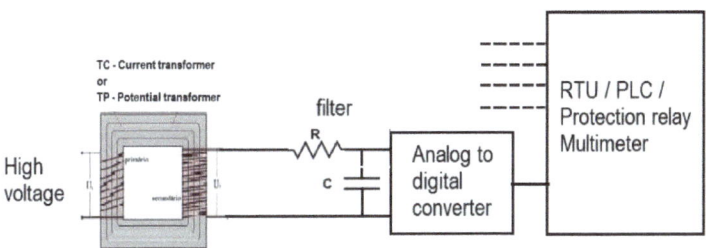

Conversion processing

Note that the evolution of electrical quantities such as voltage and current is sinusoidal. The measured value is neither the instantaneous value nor the peak value, but the effective one called the RMS value. It is approximately 70% of the sine wave peak value. See curve below.

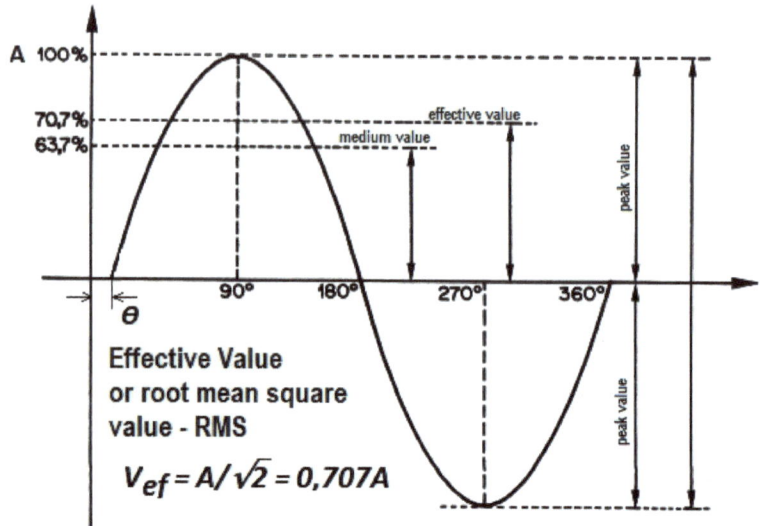

Effective RMS sinusoidal measurement calculation

    The measurement error depends on some factors of the quality and precision of the voltage or current lowering transformers, TP and CT currents. The following figure illustrates a set of TPs and TCs.

TC – current and TP – voltage (potential)

Aging can be a concern. Another source of error is the electrical circuit before the analog-to-digital converter, usually a "low-pass" type RC filter, as shown in the figure. Precision, more dependent on resistance than on capacitor, may compromise measurement accuracy. Typically components with 5% error are used. A third error in the measurement is due to the number of bits used by the digital-analog converter. Frequently, 12-bit converters produce numbers from 0 to 1111.1111.1111 in binary notation or between 0 and 4095 in decimal notation.

Thus, for example, for 230,000 volts, each bit corresponds to 230,000 / 4095 = 56.16 V. The digital converter introduces a maximum error of 1.78 % or a 56.16-volt error. That is a 56 V error for each bit of the A / D converter. Or 0.024%, which is not such an error. For 8-bit converters, a number varies between 0000.0000 to 1111.1111 or 0 to 255 in decimal notation. This error, at 230 KV becomes 901.96 volts, an appreciable value close to 1 KV. In percentage, this error is 0.39 %. It is still a tolerable error for steady-state operation.

The full-scale error method avoids processing unreliable valued variables. In pointer gauges, this error occurred when a value exceeded the end of the scale, and the pointer kept stuck to the maximum or minimum of the scale meter. To achieve this A/D converter is set to work only within a reasonable range. So, for example, assuming that the magnitude to be measured is 69,000 volts the converter is set so that the maximum value offered by the converter is well above 69,000. Typically, a 69 KV line will never exceed 73,000 V. Before that, the overvoltage protection relay would turn off the line.

On the other side of the scale, in a 69,000 V line, reaching a value less than 65,000 volts (these values are just an example and are not the actual maximum and minimum values of a 69 KV line). Before that, the under-voltage protection would have acted to disconnect the line. Thus the converter is set to convert from 65,000 to 72,000 values below and above nominal value. When it happens, we have a conversion error, and the variable is then considered invalid. The control center computer cannot consider this variable. Another advantage of using this procedure is that with a 12-bit D/A converter, the error would drop to 0.02%.

Processing

Not only sampled variable value should be transmitted to the control, as we should have already observed. It is necessary that the messages that carry these values also inform some other data about this variable. A variable identifier that this value represents must be transmitted. We append a number, the address of the variable, to the message containing the variable value. The Control center will know how to associate this number, or address, to the convenient variable. We append a flag to tell if this variable value is within the allowed range or if the variable is invalid. More recently, due to the exponential growth of the processing capacity of the microprocessors, even the instant of sampling with the GPS precision may be transmitted to the control center. The advent of the synchronous phasor measurement, to be discussed in more detail further away, came to allow this facility. Thus, an information frame containing the value of a variable also has several other data or associated information besides its binary value.

What does a control center do then?

The variables then, one by one, as they are received in the control center, are submit to the following treatments:

Volatile memory stores variables.

Usually, analog variables are stored in volatile memory as they are received. Given the number of variables coming from the field, the control center cannot save these variables in data banks. They are saved in databases organized and reliably distributed among the various nodes in the local network. Their values, their parameters already discussed, such as the invalidity flag, are also stored. An accompanying identity here determines where a variable and its components reside in memory.

Variables are offered to be seen by the operators

Graphic specialized animation programs present variables on screens as single-line drawing programs, curves on terminals, or graphic registers. They are used to consult the values of each variable available "browsing" memory.

Variables are stored in classical SQL databases

Variables are then stored and consulted using a classical SQL query language. Nevertheless, the storing period is usually well above those of sampling. It is the responsibility of specialized programs that create an analogical history of the electrical system variables. Oracle databases, PostgreSQL, MySQL, for example, are used by these functional programs.

They are transmitted to other control centers

As they arrive, the data collecting program immediately sends them to specialized communication protocol programs to guarantee they will arrive in a good state in other control centers. We will discuss these communication protocols further on.

Proprieties of an analog variable

Dozens of properties are assigned to analog variables. These properties are updated as they arrive from the field.

These proprieties, text, real or binary value, are attributed to each variable. That increases the number of information and influences the quality of the operation of the electrical system. The number of properties varies for each control center. Below we describe some of the most found properties.

Operational limits – Generally, operating limits are assigned to each analog variable, since they represent equipment-related quantities such as a voltage at a substation bus, a current in the primary winding of a transformer, etc. These limits are usually informed by the equipment manufacturer itself or dictated by operating experience or because, although they may not compromise any specific equipment, this property or condition may compromise the behavior of the power system as a whole. For example, high voltage on a substation bar does not damage the substation bus bar but compromises the stability of the electrical system. Upper and lower operating limits are usually assigned if required. Up to three upper and three lower limits are often assigned to each analog variable and can express the degree of severity that the limit value if reached, can represent. The example below illustrates the usage defined for a pair of upper and lower limits assigned to a voltage variable in a transmission line.

Warning superior limit – a value that, if reached, represents the risk that the electrical system is heading towards a situation of instability or the equipment itself may be in danger. This value, if reached, should put the electrical system operators on alert.

Urgency superior limit – A value that, if reached, indicates that some action must be taken to reduce the voltage value, as the risk of instability is imminent or equipment is near-real danger. Operators should talk to each other, examine operational guidelines or simply use their experience to decide what action or actions should be taken to adjust the voltage value to a normal value.

Operative interpretation at each limit thus depends on the type of variable associated with it. Of course, the values of these limits are static data or metadata. They are reported to the source database before the system is launched or by the operator when it seems appropriate from an operational point of view.

The warning to the operator of exceeding an operating limit usually occurs in several ways. An audible alarm can be produced by SCADA, by assigning a distinct tone to each alarm representing the severity of the problem. A warning line is generated in the alarm files and SOE describing the problem.

The operator may access this file using an application available in his operation console. The alarm is presented inside a display single-line drawing in color form. This variable value may change, in the display, to a color that indicates that it has exceeded the upper urgency limit. The yellow color is frequently used to indicate a not so drastic, alarm upper (or lower) limit attained. The red color indicates an urgent upper (or lower) limit alarm.

Acknowledgment of the generated audible alarm may be required or, it may have a short-lived duration and automatically shut down without operator action.

Hysteresis - is a value established between around the operational limits of each variable such that once this limit is exceeded, SCADA ceases to deal with it.

The figure below illustrates the treatment performed by a SCADA as part of the treatment of each analog variable. It is common to assign impossible values to upper or lower limit values whose exceeding does not entail operative risk. For example, a low winding temperature is beneficial to a transformer in hot climate regions.

Often a SCADA has up to three upper limits and three lower limits beyond the hysteresis value.

Due to the reasonably large number of variables in a supervised electrical system, those responsible should carry out exhaustive work to select the limit parameter values and hysteresis for each variable. Those in charge shall have a deep understanding of the electrical properties and operational limits of each associated equipment..

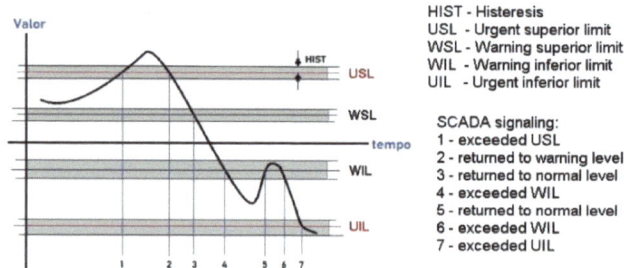

A variable evolution treatment concerning its limits

For systems that have been in operation for a while, where behavioral data have been stored already, an "a priori" auto-completion proposal considering that the phenomena approach a normal distribution is proposed (Martins, 2016).

Analog variables logic proprieties

Multiple properties with only two states can be associated with each analog variable. These states, usually grouped in a bit stream, are usually dynamic states populated by SCADA as part of analog variables handling. For example, some logical properties used in control centers are:

- *Invalidity Bit* indicates whether the variable has a reliable value or not.
- Causes of invalidity bits are bits the cause of the invalidity. For example, sensor defect, non-operative communication, variable never received at the control center, etc.
- Bits indicating that a variable not having been received from the field. It has a value set by the operator. These are known as substitution variables.

Digital supervision

The second type of measurement processed in a control center is the digital or binary variable. Just two values are needed to represent it, one and zero, 1 and 0, each representing the state of a switch, a circuit breaker, the on/off state of a pump, the actuation state of a protection relay, actuated/disengaged, etc. They are the logical variables. In the American continent the value of the variable "1" is used to represent "closed/actuated/on ..." and a zero value, "0", for "open/off/not actuated/stopped... ".

Unlike analog quantities, however, the logic variables are captured by the RTUs, PLCs, or protection relays at the instant they occur. Thus the variation of a logical input is usually associated with the interruption system of the acquisition instrument in order to capture not only its new value but also the instant that this change occurred. The previous historical discussion of data logging or SOE already illustrated that the chronology of events and alarms in a facility is the basic tool of analysis of protection experts. Thus, one more parameter needs to be added to a logical variable. The occurrence instant is always part of the frame sent to the control center.

Events chronology

It is where a new problem arises for the control centers. How to keep chronology between events occurring in different installations? The geographical distance between substations and power plants can easily reach hundreds of kilometers. How to synchronize acquisition instruments such as RTUs, PLCs, relays, etc., among different facilities? This problem remained pending until the advent of the GPS - Global Positioning System. GPSs. Even older ones use satellites to provide geographic coordinates, and altitude also provided a synchronization pulse with every second change at Greenwich meridian. It is the UTC timing.

Although messages carrying logical information do not necessarily come in chronological sequence because of the inevitable delays in communication equipment that are never uniform nor constant, the actual chronology of events and alarms, convenient processing recovers them. The chronological ordering of alarms and events must occur mandatorily in a control center.

Measurement error and security

The state of some equipment represented by logical variables not always reflects an electric state but a mechanical state. A field breaker or switch positioning, either, open or closed, is reported to the control center by the position of solenoid relays. Given the importance of these assets, a failure caused by a non-reliable value implies consequences for both the substation assets and eventual human loss.

So, it is usual to associate two complementary logic variables to represent the open and closed states for field breakers and switches. If a disconnect switch or breaker opens, the complementary logic variables associate show the value 01, where "0" indicates that the closed limit switch relay is off and "1", that the open limit switch closed, and vice versa. So, if these two variables are "11" or "00", it means that the breaker or disconnecting switch is either moving or locked. It is not either open or closed. An established timing period tolerates a maximum maneuver time, and when exceeded, we consider the variable is invalid. The invalid state means that the switch is maneuvering or breaks blocked halfway.

As for the logical quantities, the logic quantities undergo the following describing processing.

Logical variables are stored in memory

Logical variables are stored in volatile memory as they are received. Given the enormous number of variables coming from the field, the control center cannot save these variables in disk databases, but in memory-resident databases reliably distributed among the various nodes of the local network.

The invalidity flag and the occurrence time are both saved. Usually, the accuracy of ± 1 millisecond is sufficient for protection analysis.

Other parameters are also used. They are discussed further ahead. A reasonable relation between the number of analog variables and the number of logic variables in an electric power system is on the order of 10 times. For a set of 10,000 analog variables, it is common to find 100,000 supervised logic variables.

Logical variables are offered to operators

Specialized graphic animation programs put them on screens, tables, lists of events, and alarms in text files are made available for consultation, etc.

They are stored in disk SQL databases

Stored variables are also sent to database update specialized programs for structured storage. Oracle, PostgreSQL, MySQL databases, etc., are used for storing logical data. In this case, sampling is no longer periodic but occurrence time is stored. The number of rows in tables in SQL databases can grow greatly during an electric system occurrence or during the initial SCADA launching. Fortunately, logic variables do not change state so often.

Logical variables are transmitted to other control centers

At the time of each variable arrival, the SCADA data collection function immediately provides the delivery of the received values to other functions to retransmit them to other operating centers. Logical messages necessarily carry the occurrence time in order to maintain the chronology between the various centers of operation.

In some control centers, one can still find a third numerical representation or type of measurement. An integer representing the value of an energy meter such as those found in our home energy meters. These quantities are usually associated with circular and non-negative integers, just like our home energy meters.

Command functions for the electric power system

In the acronym SCADA letter, C is of control. A control center helps operators to govern the power system under their responsibility. Control centers emit orders for reconfiguration and adjustments to the electrical system. The term command here is confused with the term remote control. Remote command is a remote control.

Basically three types of command are used:

Logical commands

These are the configuration or reconfiguration commands. It addresses the disconnecting switches during maintenance actions and, more often, they address substation and power plant circuit breakers. A logic command can also reset locking relays, which guarantee safety and are used as a strategy to protect the electrical system. Some companies delegate the remote control switch only to the operators of the local control center where the maintenance actions are carried out. This increases safety except, of course, when they are unassisted.

Analog commands

They are rarer commands to be found in the electrical system. They are used to establish energy levels of operation. For example, they are used to establish a power generation base in a plant. They are found, for instance, in the CAG - automatic generation control. Reactive adjustment commands on equipment such as synchronous compensators sometimes use pulses of width proportional to the increase or decrease of reactive power offered in the installation.

TAP changer commands

Transformer TAPs are multi-position electromechanical switches that adjust the number of coil turns on the transformer's primary side to obtain a different transformation ratio. For example, a 69 to 230 KV transformer can change the ratio from 69 → 230 to 69 → 232 KV by changing the position of the TAP. TAP commands can be analog but usually are logical pulse commands given in pairs. A logic pulse command increases the voltage in the secondary, and a logic pulse command decreases the voltage in the secondary. Sometimes, the analog TAP command is translated by the equipment in the substation into a square wave of fixed duration. The command is sent as a pulse to increase or decrease TAP positioning, increasing or decreasing the corresponding transformer output voltage.

A hierarchy is established between the operators of the several interconnected control centers to divide the control actions. For example, an operator of a local control center does not perform circuit breaker maneuvers in the field. They can only observe them and check if they are performed by the control center operator, except in emergencies or loss of communication. Likewise, the adjustment actions of voltage regulation equipment are usually performed by the area or regional control centers. They have a broad system perception as they observe several substations simultaneously, and the energy evolution of the entire region makes decisions easier. In Brazil, global voltage control and energy flows between companies are coordinated by the National Operator. Although they do not act directly on transformer TAP, these commands are delegated to utilities.

In some companies, the remote control of disconnecting switches is under the responsibility of the operator of the local control center of the substation since maneuvers usually occur when maintenance work takes place in the field. The disconnecting switches are basically electrical insulation equipment to avoid discharge risks due to the high voltages involved. It is the local operator's responsibility for field maintenance work, and it is up to him to coordinates it. Local operators do switchgear maneuvers while the regional or system operators are responsible for the circuit breaker's control.

As we already know, we use two logical variables to increase the security in the observation of the positioning of switches and breakers. Isolating switches are large equipment. They are heavy ones, up to the order of five to six meters in size, and still not frequently used. Therefore, much more subject to aging. The local operator may, depending on the position of the control room concerning the yard, visually observe its state and check if the key did not lock in middle course. Very few analog commands are used. For example, they are used to establish generation levels to plants. Even tap changing commands are digital and converted in the installation to a pulse.

Photovoltaic arc during the opening of a disconneting key

The above figure illustrates the attempt to close a disconnecting switch when one of the terminals is energized.

The importance of TCP/IP in remote command efficiency

The use of optic fibers for the interconnection between facilities and the operating centers allowed a more secure reconfiguration maneuver of the substations. Generators startup and shutdown, and system restoration during severe losses can now be made from far away control centers. The most commonly used communication protocol for making these connections nowadays is TCP/IP. It is known that this protocol is highly robust concerning the reliability of the information it carries so, the possibility of sending a remote control to an element in the field and not reaching it is practically ruled out. It would be as if I received an email not addressed to me. On the other hand, TCP/IP does not worry about communication delay at all.

TCP/IP is designed for security, availability, traffic alternatives, and so on. However, we must ensure a high transmission speed or wide bandwidth. A remote command sent must reach the element in a comfortable time. The operator must not wait for it. For circuit breakers commands, operation times are of the order of milliseconds.

For the remote control disconnection switch, a time delay is inevitable. Switch disconnection is only performed very frequently during system restoration. Switches are maneuvered only in cases requiring more detailed preparation of restoration alternatives.

In any case, it is highly advisable to keep the electrical network communication system separate from the network where the company's administrative data flows. Thus, since TCP/IP guarantees that the correct message reaches the target but does not guarantee a maximum time for this to occur, as this time is associated with bandwidth, it is prudent to leave the communication channels dedicated to operation separate from all others. This is also a guarantee of harming cyber invasions.

It is not uncommon to find messages associated with sending remote control and receiving a response and associated protection that use lower layers of TCP/IP such as the network layer to reduce the traffic time of their messages.

In local control centers, the messages exchange with the equipment in the field, responsible for the acquisition and equipment commands such as PLCs, multi-meters, Digital Relays, and RTUs, is usually performed by proprietary communication protocols and very often in pairs of wires communication means, dedicated optical fibers, etc. These proprietary protocols are not always well documented and not always standardized, must be known by the computer equipment that makes up the control center. Thus, for example, the SCADA SAGE manufacturing, Cepel, supports the protocols listed below. Some are standardized, others are proprietors.

- **IEC/61850**

Nowadays, the interoperability of computer equipment that supervises and controls a power grid is only guaranteed by standardized protocols. The IEC 61850 protocol is a protocol that guarantees interoperability. In addition to being a modern object-oriented communication protocol that simplifies its implementation, this protocol also offers the use of messages called GOOSE (Generic Object Oriented Substation Event) which, without the need for additional and specific wiring, provides a high-speed communication, even under the TCP / IP protocol. GOOSE uses OSI application layer number 2, avoiding other layers and thus allowing, for example, high-speed circuit breaker relay trip commands. Systemic protection could thus, at some limit, be contemplated from local or even regional control centers.

- **TASE.2 / ICCP-MMS**

TASE.2 is a protocol designed under the assumptions of safety, low component and installation cost, interoperability, training costs, vendor-available support, and general-purpose industrial solutions. TASE, 2, part of the Utility Communications Architecture - UCA, is described in three documents: IEC 60870-6-503 for services and protocol, 802 for object models, and 702 for application profile. TASE.2 standard is also known as "Inter-Control Center Communications Protocol" - ICCP. It begins to be used heavily as a communication protocol between operating centers, whether the local, area, regional, or country-wide in the electricity industry. Cepel's SCADA SAGE includes blocks 1, 2, 3, 4, 5, and 7 - ICCP protocol and native MMS library.

- **OPC UA (IEC 62541)**

There are currently over 22,000 market products offered by over 3,500 companies that provide this communication protocol. It is a service-oriented, not object-oriented protocol. SCADA EMS SAGE provides server and client for external applications and access to objects and variables of recommended OPC models.

- **IEEE C37.118**

This is protocol has recently been widely used for efficient acquisition and distribution of synchronized phasor measurement data, to be discussed later. It is usually built-in as a service in phasor measurement units, Phase Meter Unit (PMU) as a metering server, and should be present at control centers as clients. It uses TCP services for initial connection dialogs and the faster, anonymous UPD service for spontaneous data transmission. It is a modern protocol that uses a GPS connection to conveniently and synchronously date analog voltage values along with their values and phase angles.

The phase angle, as will be discussed in more detail below, is fundamental information. When compared at various points, indicates in advance the proximity of the electrical power system to an unstable region. Other quantities are also acquired by a typical PMU, taking advantage of its modern and precise analog-to-digital converters, such as the grid frequency with reasonable accuracy. A more detailed description of synchronous phasor measurement and its connection to operating centers will be described later in this document.

- **IEC 60870-5-101**

This protocol is better known as the "101 protocol". It is of wide use in the data transmission of electrical power systems. It is a standard created in the 1990s and widely used in the Electrical Sector. It is a serial protocol in RS232 used for low-speed transmission (from 9600 BPS to the fantastic 64,000 BPS today) in balanced and unbalanced mode.

- **IEC 60870-5-103**

The same family as above but specialized in communication with protective equipment.

- **IEC 60870-5-104**

Same as above, but allows using the TCP / IP protocol as a communication transport.

- **DNP V3.0**

The DNP protocol was designed for communication with substations and other electronic devices. It supports remote control applications and is widely used today by the power industry in the United States. Originally from Harris, it is today maintained, updated, and supported by a user group, DNP User Group.

- **MODBUS**

Originally developed by Modicom in the 1980s, because of its simplicity, it became a de-facto standard. It supports both serial and TCP-IP communication. It offers two encodings for data, RTU when data is encoded in binary, and ASCII when encoded in ASCII. The late is more readable for those who love to read messages that travel between two devices.

- **SNMP**

Its acronym comes from "Simple Network Management Protocol." It is not used as a protocol for coding electrical data but is widely used for supervision and control of equipment and computers that control the electrical system such as routers, switches, workstations, desktops, etc.

Other less commonly found protocols are CONITEL C-3200, MICROLAB-STD, ALTUS AL-1000, Leeds & Northrup LN-57, and Allen Bradley 1771.

It is then necessary that control centers have deployed and offered access to a huge range of communication protocols to address the Babel of equipment from various manufacturers, measuring instruments, protective equipment, meters, and remote terminal units.

To overcome different manufacturers protocol standard interpretation there are documents known as User Companion, or as Profiles, where each manufacturer indicates which aspects and messages his computer or equipment attend.

This simplifies discussions about the difficulties of interpreting the various messages and features described in the usually lengthy full documentation of each protocol when two companies begin attempts to establish communication between their equipment..

Systemic protection

Quality of communication has been so enhanced by optical fibers that companies already use specific programs, including artificial language, to perform the automatic restoration of electrical systems covering large regions. Systemic protection functions are already in place today. Functions disseminated by several control centers which communicate using strategy to implement automatic restoration, not more than in one installation but a group of them expanding the traditional concept of protection of the equipment or the installation for systemic protection.

Supervisory services offered by a SCADA

Now we start to confuse the term electric power system control center with SCADA. A control center is usually a SCADA. The supervisory functions provided by SCADA are described below.

All SCADAs have a set of essential basic functions. There are variations in techniques and offered functions, described below. They vary for each SCADA supplier.

It is necessary to organize information, logical data, and analog data. SCADA functions are responsible for that. That is why the document describing these functions is called "Functional Analysis."

This document describes which functions will be available in the SCADA, the SCADA computer topology, how many and which services will be allowed, which variables and their attributes will be treated for each installation, which protocols will be used for each of the facilities, etc. In other words, it is necessary to parameterize the SCADA before putting it into operation.

Static data or metadata and dynamic data

Two types of databases are typically generated for the definition of a control center. An offline parameterization database and a dynamic online database derived from the first. The latter, predominantly composed of the information of the electrical system and its evolution. Parameterization data or static data are also called metadata. On the other hand, dynamic data is the information arrived treated, stored, and processed by the SCADA during its operation. Variable values frequently change inside the database.

It is necessary to describe in detail all the static characteristics of the SCADA for a given control center. During SCADA launching, its functions read table parameters, previously prepared by SCADA's specialists to fit SCADA to a specific situation. As soon as it finishes launching, it begins to talk to the installations, other computers, and devices and receive data to be processed.

Alternatively, in more open systems from a computer point of view, that is, systems where the manufacturer does not hide the source files of the programs, these tables are link-edited to the source programs to produce the set of SCADA executable. The use of a SCADA then requires two steps. An off-line parameterization step and an online operation step, when programs performing the various functions are neatly launched automatically or manually by IT specialists.

The notion of priority of execution of the various computer functions that make up an operating center is a feature that makes it difficult to use with operating systems where the establishment of an execution hierarchy between the tasks is not clearly offered by the OS. Microsoft's traditional operating systems, for example. So, much of the control center software lives under the mantle of Linux systems or even more specialized ones like the Canadian QNX, called "Linux like."

Data storage

As discussed earlier, data received from the facility is stored first and quickly in a memory database and spread through a computer LAN, and later in traditional SQL databases for storing the historical evolution of the electrical system.

The in-memory database is always being changed by field events. The disk database, in turn, grows with each arrival of a logical field variable and periodically by querying the database access function to the resident database.

Data logging – alarm and event storage

As previously discussed, the data received from the installations are quickly stored in an in-memory database on computer LAN. Later on, they may be stored in traditional SQL databases to keep the historical evolution of the electrical system.

The in-memory database keeps being altered by field events. As each analog or logical variable arrives from the installations, the SQL disk database grows. This database is consulted periodically by other functions.

Data logging - SOE history

The list below illustrates a data logging segment of a few minutes in an operation center. This list is produced in a text file, and its meaning is natural to the operators and analysts of electrical systems. It is an example of the already mentioned SOE - "sequential of events." In the example below, the file only contains the instant of occurrence of each event. The date should probably be part of the file name. Briefly, for example, the second line of this file can be read as:

```
00:08:11.354  SDMR3456KL  ACTUATED   SDMR3456KL 86 Protection
00:19:12.568  SDDJ704POS  OPENED     SDDJ7-04 Position Breaker
01:19:12.568  GUA6KV001   ULT LIM    GUA6KV001 Guama 69 KV bus 1
02:19:12.568  GUA6KV002   INVALID    GUA6KV002 Guama 69 KV bus 2
```

These daily, weekly or, monthly files and/or tables can reach several lines that can exceed thousands in the supervision of electrical systems of approximately 100 to 200 thousand variables. It is common to find a line repeating itself hundreds of times when an old or defective field relay has its contact bounces. This suggests that the files or tables should be cleaned, usually known as the "data warehouse." This information can be available as CSV files, Excell spreadsheets, and even PDF.

Storage of electrical quantities - analog history

Today, all control centers store electrical quantities collected at the facility and reach the control center. Their values, the instant of arrival in the center, installation collection instant, its attributes, and value are all stored. Dating accuracy that accompanies value and its attributes in the transmission packet from the facility to the operation center is usually not the same as the one used for the logical data. Milliseconds of sampling accuracy are only found in modern data acquisition equipment, the PMU (Phasor Meter Unit).

Analog measurement storage – analog historic

Today, all control centers store electrical quantities collected at the facility and reach the control center. Their values, the instant of arrival in the center, installation collection instant, its attributes, and value are all stored. Dating accuracy that accompanies value and its attributes in the transmission packet from the facility to the operation center is usually not the same as the one used for the logical data. Milliseconds of sampling accuracy are only found in modern data acquisition equipment, the PMU (Phasor Meter Unit).

Analog history in text files or spreadsheets

The analogical history of the electric power system is usually stored in the form of text or spreadsheets. CSV files (Comma Separated Values) or even proprietary formats, suitable to be imported by spreadsheet programs. More recently, analogical histories are found in pure text files in XML (extended markup language), JavaScript Object Notation (JSON) files, or even more rarely in ready-made spreadsheets.

Often database files are stratified into periods. Its name identifies the time interval contained in the samples stored inside. For example, the file name BD_2018-11-10.csv indicates in the text itself where its contents are filled with analogical information of the electrical quantities stored in the database BD by the operation center on 2018/11/.

Often, database files are stratified into periods. Its name identifies the time interval contained in the samples stored inside. For example, the file name BD_2018-11-10.csv indicates in the text itself where its contents are filled with analogical information of the electrical quantities stored in the database BD by the operation center on 2018/11/.

History stored in SQL databases

It is common to associate the adopted tag or mnemonic to simplify the description of the variables in a control center with an integer. For example, tag GMKVBR601 associates the full text of the variable, something like "Bus bar Voltage Number 1 of the 230 KV yard of Guama Substation". This facilitates SQL access to variables. Usually, a number, called index, is associated with the TAG. This procedure uses the fact that computers find numbers easily than texts.

Thus, in a first approximation, one can, for example, associate a number or index in the alphabetically increasing order of the TAGs used. For the TAG example mentioned above, GMKVBR601 could be associated with the number 1237 if, for example, it is the 1237th variable in the alphabetical order of the enterprise variables. If this association strategy is adopted whenever the number of variables is enlarged, the number associated with the variable will probably not be the same.

This will cause a problem retrieving past information from this variable if this number is used in the historical database. In our not polarized understanding, generated hash algorithms are a better choice if we want to maintain uniqueness between the index and the TAG of each variable to be stored in a historical database.

The amount of data stored - data warehouse.

Today we find control centers monitoring around 30 to 50 installations. They are fed by more than 100,000 variables, usually with 10 to 20% of analog variables and 80 to 90% of logic variables. The latter, despite the majority, do not, fortunately, arrive very frequently in the operation center. Analogical variables are almost cyclic and could fill hundreds or thousands of entries in text files or database tables and therefore are not often stored on disk on arrival but periodically by reading the memory database.

The need for file cleanup processing is therefore imperative in control centers. This activity, known to data processing staff as a data warehouse, should be used as a permanent process in operation centers. In addition to the cleaning described above, more subtle cleanings are also needed. Thus artificial intelligence - AI and its algorithms are often used by companies. These modern functions, however, are not embarked frequently in supervisory and control systems delivered by manufacturers. Electric companies should seek the means to promote the cleanliness of their databases.

Time synchronization between installations – GPS

As discussed earlier, the analysis of a sequence of events does not depend on the precise sampling time but on the chronology between events, the order they occur. An installation event set supervised only by one UTR, even the oldest ones, guarantees sufficient time accuracy between events because timing precision ensure. For example, an interval of one-thousandth of a second is enough to achieve a good timing accuracy. The chronology problem appears when the logging data arrives in an operating center from several facilities far away from the operation center.

This problem remained unsolved until the arrival of geo localization technology. The GPS - Global Positioning System device, available today even in our cell phones, gives us latitude, longitude, altitude, and eventually speed. It also produces a synchronization pulse at each minute change. Of these GPS characteristics, the latter one, has solved the synchronization problem between the supervision equipment of faraway facilities and the control center.

Specific and adequate GPS equipment in substations and power plants uses, for example, a low-level protocol such as IRIG-B, which, although expensive, offers a precise synchronization between computer equipment (Protocols applied for time synchronization in a digital substation automation. , 2018).

The installation of GPS in each facility then became a must to ensure correct supervision of the electrical data. A lower-cost solution is offered today when the TCP/IP protocol is used in the communication medium between the operation center and its facilities through a TCP/IP sub-product, the "*network time protocol*"– NTP (Wikipedia, 2018) Operating state presentation in real time.

Using still the fact that operating centers are the operator's eyes and ears, a supervisory system is necessarily able to present data obtained and stored in human-readable and comfortable forms. There is an expertise area that studies the best men-machines relationships that leads to a comfortable, safe, and non-detrimental manner. Ergonomics is the name of this discipline. Ergonomics studies the work organization in which there are interactions between humans and machines.

It is thus necessary to present the topological state of the electrical system in a familiar way to technicians, such as diagrams and electrical circuits. Dynamic data representing electric quantities shall be expressed as numbers and, in the case of historical data, as curves where the behavior can be humanly visualized and behaviors compared by the operators and analysts.

Single-line diagram – the operator´s eyes

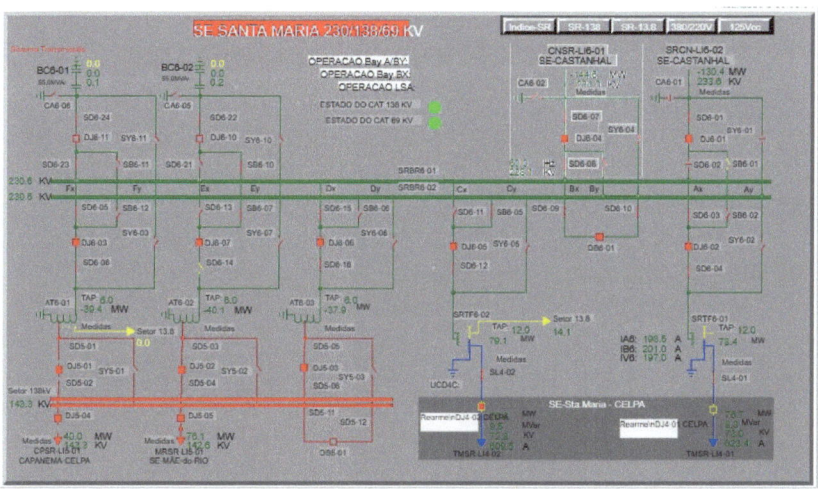

The single-line diagrams would then be the operator´s eyes of an electric power system. Much of the time an operator spends in the control center is observing single-line diagrams. A single-line diagram represents only one of the phases of the electrical system for simplification, as their values are almost identical in steady-state. The phase drawn in operators display is usually the B phase. The figure below illustrates a typical display showing a single-line diagram of a substation.

Installation topological drawing

The electric circuit shows the substation bars, transmission lines, and installation equipment. They are symbolically represented by colored lines and iconic coil designs for reactors and transformers. These symbols are already familiar to technicians and electricians. Colors represent the associated voltage class according to the company's choice. For example, the figure above represents a substation that transports electrical energy at 69 KV voltages, blue color, 138 KV, red color, and 230 KV, green color. The substation bus bars are represented by thicker lines.

So far, we have presented the static part of the screen as a single-line diagram. The electrical diagram, the equipment like reactors, capacitors, transformers, lines, and bars of an installation are static representations. They do not change with changes in the electrical system variables.

Analog variables representation

Analog quantities are represented by real numbers, with accuracy compatible associated metering system and suitable for fast human identification. The operator is more concerned with the order of magnitude than its exact value up to the nth decimal case. In the example above, the green analog variables represent a reliable real variable value. A variable in yellow represents an invalid variable or invalidated by the measurement system. Other colors may also be associated with the representation of analog variables indicating that the recommended operating limits are exceeded. Today's computer graphics offers several alternatives for the presentation of analog variables. Even pointer display can be emulated at a video terminal to attend to nostalgic needs or when pointer positioning relative to the meter scale is paramount..

Logical variables

Logical variables represent the state of switches or equipment such as circuit breakers, disconnecting switches mainly and, in addition, any other on/off type equipment such as pumps, protection relays, or protection instruments states are represented by the interruption of lines. For example, in the figure above, the circuit breakers are represented by empty squares when opened and filled squares when closed. Switching switches are represented by classic switch designs. These representations can also be associated with colors that represent the equipment's validity status. For example, in the figure, the DJ4-01 Celpa circuit breaker in the lower right-hand corner of the 69 KV yard (in blue) is yellow painted to emphasize that its state, represented in the diagram as open, is not at a reliable state. This can be confirmed by noting the presence of flux through the TMSR-LI4-01 line.

All these representations are dynamic ones. They change in real-time with the change of the associated measure in the field.

A control center supervisory system is perfect to represent the system's steady-state. Transitory events, when abrupt variations of electric behavior occur, are naturally filtered by system sampling rates. 2, 8 10 seconds sampling rates are used for quantities, such as frequency, voltage, and current. Sampling rates up to minutes may be used for temperature, water level, etc. It is important to note that there is really a compromise between the amount of processing data, the sampling interval, and what digital technology offers nowadays. It would be necessary to use much faster sampling to observe the transient regime of an electric power system. Thus, a SCADA presents the behavior of the electrical system in a steady state.

Audible alarms – the operator's ears

The operator's ears are audible alarms associated with variations in magnitudes that indicate that the system is moving from a comfortable state of operation to a state of risk of electrical instability or that some equipment has changed its state, such as a circuit breaker opening. Supervisory systems offer the possibility of associating audible alarms with different tones. Thus, for example, it is possible to associate audible alarms with the circuit-breaker opening. The disconnecting switches are not frequently related to audible alarms, as they are only used during installation maintenance maneuvers. They are always operated by an operator's command. It is assumed that he knows what he does. Protection relay actuations, in turn, are usually associated with emergency alarms.

Historical graphs

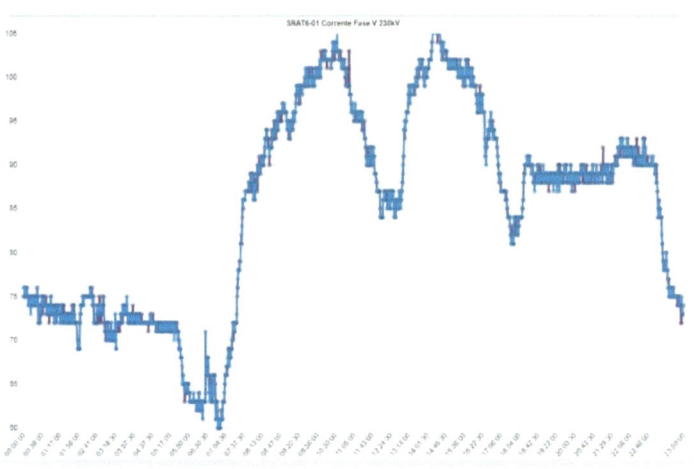

Time evolution of a transformer current

The figure above represents one of the courtesies often offered by supervisory and control systems in control rooms. The historical graph of an analog variable associated with an electrical installation variable magnitude. In this case, it represents the daily variation of the current of phase V of a transformer electric current in a substation.

This tool is not used very often by the electrical system operator and much more by the post-operation analysis engineers to evaluate electrical behavior after a system contingency.

It is usual to use a real-time graphical representation of variables. They are called trend graphics and show continuous variable evolution. This type of chart, usually offered in control centers, is called a "trend graph." When the operator suspects strange behaviors of an electric variable, he may place it under his more closely watch. Several operation centers use a visible trend graph, for example, showing the deviation of the frequency value from its nominal value, 50 or 60 Hz. This deviation gives a concrete idea of the unbalance between available electric generation and consumer demand. A frequency value greater than nominal means that there is generation left over. A frequency value less nominal value means that the load is not being fully met.

SOE on line consultation

Another usual tool available in control centers is the online or deferred query tool of sequential event and alarm registration. In many control rooms, a dynamic display of the last occurrences of "open/close"; "alarm/normal"; "operated/not operated"; etc., of the various equipment and relays in the field, conveniently ordered and chronologically dated by order of the arrival in the control center. It is by these "eyes" that the operator follows the logical evolution of the electrical system - (Marinelli, 2017).

Automatic voltage control - AVC

One systemic control that the companies started to use, as it did not require fast remote commands, was the automatic voltage control - AVC. (Corsi, 2015).

A basic strategy for AVC would be to use a timetable with a periodic schedule of the voltage levels. These tables contain a list of transformers equipped with voltage control. This table would hold a daily and hourly voltage schedule. Each substation control center with controllable transformers would then run this function. These substations distribute energy to towns or large industrial consumers.

The AVC function then watches, at every few seconds, if there is a violation of voltage levels delivered to the consumer. It does so for each transformer at the substation. It waits for a few seconds more and decides to emit a transformer command to adjust the voltage level according to the schedule table for this hour of the day.

These tables are usually results of energy furnished contracts between the transmission company, and a distribution company, and large consumers, such as industries. These tables have voltage-hourly values established for each weekday with three to four months duration. They are negotiated at the regional control center and computer transmitted to local control centers. They are known as supply contracts.

For example, a substation equipped with an AVC function at its local control center, even if unattended, would maintain its voltage control even during possible long-term communication losses. The experience in the Electric Sector has shown us how this function improves the operator's life quality, taking away the permanent concern with the verification of the normality of the voltage profile at all times.

A few months after its implementation, initial fears expired; operators from local control centers and regional operation centers believe they are no longer allowed to live without AVC as we no longer allow ourselves to live without a telephone. The figure below illustrates an example of a strategy that can be adopted for implementing Automatic Voltage Control - AVC in a regional control center and a local control center in a facility.

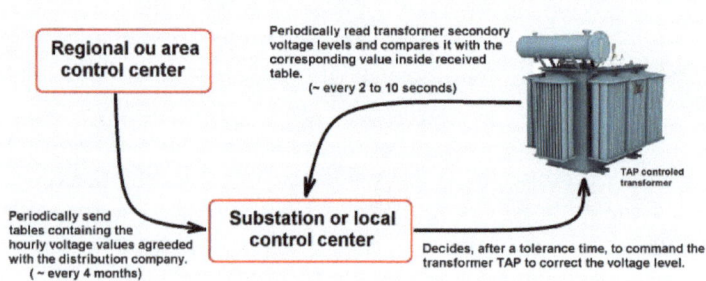

An AVC - Automatic Voltage Control strategic concept

And here we finish discussing SCADA functions.

Energy Management System - EMS

Attending government recommendations in the United States that followed the 1965 blackout, in 1981, the Brazilian Government promoted what may perhaps be termed as the first training for specialists in electrical power systems advanced analysis functions, such as state estimation, contingencies analysis, etc. These will be discussed in more detail below. The name of the training "Electrical Application Engineer" was inadequate but was one of the precursor courses that allowed the development of systems nowadays known as EMS.

Here SCADA enlarges and becomes an EMS - Energy Management System. EMS is like a new shell over the SCADA. It adds to SCADA a set of new functions that improve observation accuracy, operational advice, and prediction of what may occur shortly, thereby increasing safety.

These new services are briefly described below.

The mathematical tools and functions described below are EMS components that belong to an area of knowledge called "Power Systems Safety Analysis."

Power flow calculation – The Load Flow

Power flow calculation or "load flow" has already been used by the energy planning areas of energy companies since the mid-1960s or even earlier. It consists in mathematically modeling an electrical system when you know only part of the data. Modeling is done by approximation. From a given set of data obtained by the supervisory system, or obtained for equipment published electrical characteristic, and using Maxwell equations, determine the steady-state of the electrical system.

The term "state" is here used to denote the set of complex values of all the voltages of each substation bus, that is, its amplitudes and angles, and the active and reactive power that flows through each transmission line. This allows the verification in advance of whether the available transmission lines support the currents that will flow through them to meet the demand for planned loads. This allows, for example, to evaluate the behavior of the electrical system in case of contingencies when disconnecting lines, generators, or loads. It is the tool to determine the future behavior of the whole electrical system before it happens. The repetitive use of load flow is, therefore, a powerful and indispensable tool of energetic planning.

However, there are some problems with modeling electric power systems. Power systems are non-linear systems. The Mathematical model of a power system uses partial differential equations. Mathematical modeling of electric power systems uses large matrix calculations, which posed severe problems until some time ago when processing power was still limited.

Even knowing the active and reactive powers of the generators and loads impedances of the transmission lines and other types of equipment, even knowing transmission lines' active and reactive power, obtained by the supervision system, even knowing some of the bus bar voltages, the number of unknown variables, using a Kirchhoff formulation, are greater than the number of equations.

Another problem is that electrical systems nowadays contemplate hundreds or even thousands of substations. Thus, the size of the arrays may be unbearable for processing, depending on the computational techniques used.

The problem is not only serious because, usually, the matrices of admittances | Y | are sparse because the great majority of substations are only interconnected to few others close to them and not to all others of the system. Nevertheless, matrix | Y | contains a high number of zeros minimizing the inevitable matrix inversions problems.

The missing line means null admittance, and there are mathematical methods that can simplify calculations or at least reduce processing time, using the fact that matrices are sparse. However, the raw approach to differential matrix equations is too cumbersome to process with the increasing number of bars or number of equations.

Usually, we use successive iteration approximation methods that take into account that if an electrical system is in a steady-state, a stable situation, its bus bar voltage values are very close to their nominal value. This method will be briefly described below.

The methodology begins by reducing all quantities to a common-base or a nominal value denominated PU (per unit) so that voltages in PU are very close to one. The figure below illustrates a commonly used model in the power flow calculation problem, and the known and unknown problem values.

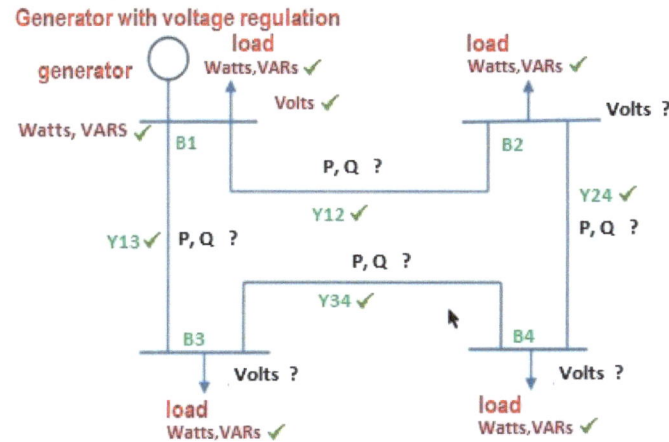

Modeling used for power flow calculation

A brief mathematical discussion of a basic approach by successive approximation used to calculate power flow for the example of the figure above can be found in (Power System Load Flow Tutorial Part 1, 2018).

With greater processing speed power available, it is now possible to perform power flow calculation, no longer in large and heavy processing centers, but even using minicomputers or desktop computers in a computer network used in a control center.

The advantage of using load flow in control centers is that the previously entered information to be used by the power flow program is now measured by SCADA. Thus, the complex power generator values, load values, and fixed voltage values by the voltage controller, in some plants, are now obtained in real-time.

In this way, two representations or two models or two "states" of the electric power system are obtained. One is determined by the calculation of the power flow, and another is obtained directly from the SCADA.

These two representations, or models, or "states," will be necessarily different. For example, transmission lines change their impedance values with age. Or, for example, the recent presence of a corn crop beneath it changes its capacitive value to a different one, obtained from supplied by the manufacturer of the line, when installed. This gives us an advantage if we use SCADA measured values to the theoretical values calculated by the power flow.

On the other hand, SCADA uses A/D analog-to-digital converters and filters, previously discussed herein. They may be out of order, and the measured values will show significant errors in addition to the errors inherent in the number of A/D converter bits. This gives an advantage to the load flow calculation concerning the previously given values obtained from manufacturers.

It is important to note that power flow calculation is an internal service offered by reputed control centers. It differs from widely available programs for purchase because it uses SCADA data to produce the state of the electrical system. Programs such as Cepel's ANAREDE, PSCAD, DigSilent, ETAP, Power Flow from Easy Power, and even Matlab and its libraries are offline programs that require us to provide topological and analog system data to produce the model. The internal program of control centers does not need this. It obtains the description of the electrical system from the SCADA online database. (Cepel, 2018)

How to reconcile the two "states" of the electric power system is then work for the state estimator described below. How to reconcile the measured state to the modeled state is a work for the state estimator function. (Power System Load Flow Tutorial Part 1, 2018)

State estimation

The state estimator is a mathematical tool or algorithmic program that uses as a starting point two large arrays. One array containing the measured state of the electrical system, and another one containing the resulting state from a previous power flow calculation for a hypothetical electric circuit resulting from the system proprieties data, the electrical characteristics of its components such as transformers, reactors, the theoretical impedance of the transmission lines installed, etc.

Naturally, it is mandatory to know the electrical system connection. We must know in advance how each piece of electrical equipment is connected to its neighbors. This is the work of another EMS chain program, the state configurator discussed in more detail below. It is mandatory to know which lines are connected, which transform ratio is offered by each transformer with TAP, which reactors and capacitor banks are connected at that instant, their reactance values. And so on.

With this data, the state estimator works, using an optimization algorithm to minimize errors. It determines a third array representing the "estimated" state of the electrical system under observation. Thus, all the electrical system measurements, voltages, active and reactive power flows, etc., are now represented by two values, a measured value and an estimated one. The latter should be used by the operators on their screens and single-line diagrams on their monitors. The complete approach from the engineering point of view can be found in (Monticelli A., 2011).

This function or service determines the best-estimated state of an electrical system from the values measured by the SCADA and those obtained by a power flow calculation of the last state. The estimation function or service is usually capable of pointing out gross errors, filtering improbable or physically impossible values, and prepare an environment for contingency analysis from real-time data. A state estimator is thus a tool for the daily use of electric system operators. The operators should use estimated values in place of measured ones. (Monticelli A. , 1999).

State configurator

State configurator is an auxiliary program or function required by the EMS but is not directly used by operators as the state estimator discussed above. The state configurator is a program to generate the electrical system topological model. As if a "system connection photo" each time any equipment changes its connection state to neighbors. This topological information is used, then, by the state estimator. It is awake by the reception of the SCADA digital data function.

It processes the state of circuit breakers and switches that connect equipment's to lines. It determines basically which bus bars are connected to reactors, capacitor banks, shunt coils reactors, etc. Each connection may be direct through a single circuit breaker or switch or through alternative logic connection schemes.

In an alternative path connection, it is necessary to do some logical AND, OR, NOT calculations to produce a ready drawing or topological map of the electric circuit.

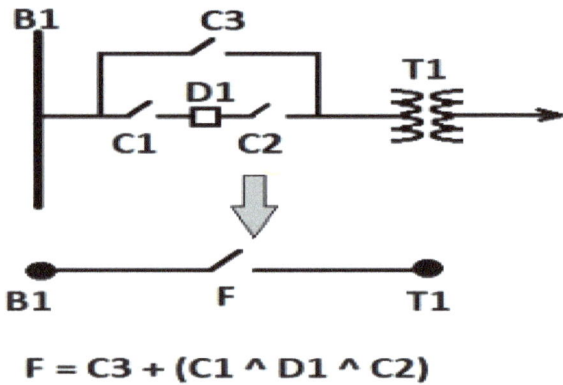

$$F = C3 + (C1 \wedge D1 \wedge C2)$$

Typical state configurator calculation

For example, the figure above illustrates how the bus bar equipment is connected to a transformer and a transmission line. The configurator performs the logic function below to decide on the connection between the bus and transformer equipment.

F = C3 + (C1 ^D1 ^C2) where,

"1" and "0", respectively, represent the closed and open states of the sectioning equipment, "^" is the logical "AND" function, and "+" is the "OR" function.

This function represents the connection state between the transformer equipment, T1, and the substation bus, equipment B1. The program that elaborates the power flow and the state estimator promptly receives a topological "drawing" of the electrical system shown in the figure that presents a classical electrical diagram.

The processing chain of these functions can be summarized in the figure below. Other strategies may well be adopted but this is a frequent one.

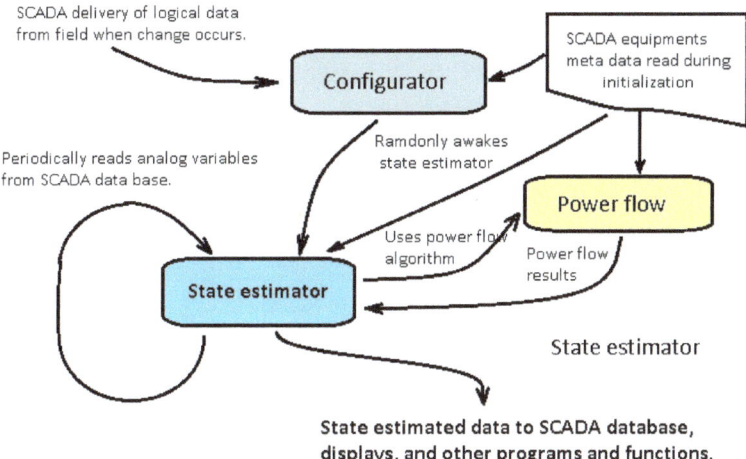

Contingency analysis

A SCADA only, despite an excellent system of observation of the state of an electrical system, is not but a damages announcer. When an alarm, whether in text form, an audible alarm, or an object "blinking" on the operator's screen, announces the loss of a transmission line, the disturbance has already happened. All we can expect for the moment is that severe damage is not done.

The contingency analysis function, in turn, allows you to answer "what if?" questions. For example, the experienced operator or a pre-dispatch engineer may, given that the system is in a state, check how it would behave if a given line were suddenly switched off. What would happen if a generator or generator group were shut down. If a sudden loss of charge occurs to a consumer. What would happen to the system when it is necessary to disconnect, for example, a reactor to put it in maintenance?

These questions are posed by electrical planning personal but not with the system alive. We have a planning system with a real-time system. Contingency analysis is the first service to prevent accidents. To this day, as far as we know, however, very little contingency analysis is found in daily use in control centers where we have been.

An electrical system is said to be safe when it is able to return to a normal situation when it undergoes a disturbance. The normal situation is that in which there is a balance between generation and load, frequency is very close to its nominal value (50 or 60 Hz), voltages in the substation bus bars are within limits electrically and contractually acceptable and currents in the transmission lines or that pass through transformers do not lead to the risk of overheating or premature aging or even damage.

The contingency analysis function can then be classified within the area of safety studies of the electrical system. Briefly, the contingency analysis function receives information from the state configurator and the state estimator to know the current state of the electrical system. Then by operator action, it determines the next state of the system if a given contingency occurs.

To achieve this, it begins with estimated values, changes topology accordingly to meet the request, and executes a load flow to obtain the new state. The contingency could be, for example, the shutdown of a line, the connection of a reactor, the loss of a load, etc.

The complexity of the problem to be solved depends on the level of the contingency submitted for analysis. If the investigation is to be made for the input or output of only one kind of equipment, the contingency is said to be level "1", also called "N-1" to represent that all equipment N is in the normal state and only one is not.

If, on the other hand, there is interest in investigating how the electrical system would behave if, for a given contingency other contingencies also occur then, the analysis would be level "2" or "N-2". This is much more complex and costly processing from a computer point of view.

From the state presented by the contingency analysis tool, specialists can repeatedly select alternatives to avoid possible damage, for example, raising voltage levels, injecting or withdrawing reactive power from control equipment, etc.

The factor that determines which contingencies are the most important to be selected at any moment is the operator experience. It is up to him to prepare the electrical system for a given analysis scenario.

It is important to note that contingency analysis is primarily a planning tool. It prevents possible system loss but also previews the system state in the case of a programmed loss, for example, the need of transformer maintenance, the insulation of a bus, etc.

An observation that must be made is that the algorithms used for power flow approximate calculation use the search of the minimum of a function, such as the descending gradient algorithms used in artificial intelligence techniques. It is to be expected that, at times, the minimum is not found or, in other words, the algorithm does not converge. This may represent physically or electrically that the electrical system has suffered a severe contingency and has entered into situations of instability or the algorithm is impure. In this case, the modeling used is no longer valid.

Therefore, the analysis of contingencies should be limited to light contingencies, those in which the system can still find a state of equilibrium after the contingency.

Traditionally, as already discussed, contingency analysis is already used in an off-line environment outside a control center. Planning scheduled shutdowns for maintenance, use contingency analysis to ensure system security. It uses power flow programs to generate the state in contingencies. They have already been used for a long time by specialists in electric planning.

Starting from a determined state of the system, it elaborates from a power flow result, using nominal values of generation and load and the electrical data of the various equipments and given an interconnection, the topology of these types of equipment.

Today's state-of-the-art control centers offer this planning tool using a "snapshot" of the state of the electrical system obtained by SCADA as part of EMS services, which undoubtedly is a great advantage.

The advantage of having this tool in the computing environment of an operation center is SCADA allows the use of the present state of the system that can be obtained by a snapshot of the state of the measurement or, better still, the estimated state of the system.

This allows for short-term planning in a pre-operation or pre-dispatch environment, thus considerably increasing electrical system safety. A brief and precise description of contingency analysis can be found at (Mishra, 2012) & (CEMIG, 2009)

Load forecasting calculations

Load forecasting predictions have already been carried out in electric power companies for many years. The need for load forecasting is directly related to the cost of energy production. Overestimating the future burden of a system implies unnecessary generation costs. Underestimating the future load may lead to the possible start-up of thermal parks, usually bringing expenses not foreseen in the budget.

At the time of publication of this book in Brazil, when this happens, companies raise "red flags" in our electric bill to signal that they are transferring unplanned expenses to us, sometimes from planning errors. This always happens when they begin to use thermal generation.

The foreseeing forecasting future is distinguished into three categories. Short-term forecast where the forecast horizon is days, medium-term forecast possibly one to two months and long-term, carried out by the areas of operation planning, where the horizon reaches months or years and uses the growth of the population, GDP growth, etc. Here the number of variables used is no longer restricted to electric observations of past demand behavior.

The classical load forecasting methodology is being adopted by energy companies uses the Box-Jenkins model (Barros, 2014). Classical statistical methods are used, such as ARIMA models (AR: autoregressive, I: integrated, MA: moving average).

Nowadays, artificial intelligence techniques offer more accurate models using, for example, artificial neural networks - RNA and support vector machine trained with past data to make future predictions.

In any prediction methodology, data used are the load values described in time series and obtained from the behavior of the loads in the last time interval. The behavior of an electric power system is strongly seasonal and almost vegetative and linearly increasing. Consumption on a day of the week is often very close to consumption on the same day of the previous week. The average consumption in the summer months is very close, slightly different from the consumption average last summer. Thus the statistical algorithms are reasonably successful for forecasting, especially those for the short and medium-term. Barros.

Economic dispatch calculations

Economic dispatch is the name given to the function that tries to promote the optimal generation of energy for each group or generator element in an electrical system. It is a specialized program that seeks to satisfy economic operational and electrical restrictions of available generation sources and the transmission capacity of associated lines.

The general idea is to attend to the loads at a minimum cost. The generators set with the lowest marginal generation costs should be used with priority.

Basically, the economic dispatch program receives information from systemic constraints informed by the operator, values of electric quantities obtained from SCADA to determine the set point for the automatic generation control - AGC. A detailed description of the economic dispatch calculation can be found on (Wikipedia, 2018).

Automatic Generation Control, AGC.

Electromechanical power generation plants of any type either thermal, wind, or hydraulic are subject to braking. There is a reduction of the speed of rotation of the generators when an additional load is applied, demanding more energy, or energy source decreases. Likewise, generating units tend to fire, that is, to increase their speed of rotation when a load is withdrawn. The greater the magnitude of this load, the greater the braking, the greater the speed decreases. For this reason, all of them have automatic primary speed control mechanisms which bring the rotation and frequency back to their nominal value.

On the other hand, as already mentioned, the frequency of the electric network is directly proportional to the speed of rotation of the generating machines. Thus the frequency of the power grid is biased whenever the generating units are subjected to load variations. There remains an inevitable frequency control error when load variations occur. The control mechanism responsible for bringing the frequency back to the default value, 50 or 60 Hz, depending on the country, is called Automatic Generation Control (AGC). The second responsibility of AGC is to keep the values of interchange energy between the companies at their points of interconnection in the values commercially contracted for each determined period.

The AGC function ultimately establishes the set-point or the base power for the generating plants periodically. Usually, this base power value is created as a percent value so that all plants receive the same set point and adjust proportionally according to their generation contribution capacity. AGC gets information on the power exchange values contracted at the various interconnection points, obtains the frequency deviations suffered by the generators during the load variations, and promotes the return of the frequency to the default value.

A last auxiliary function is only available in power plants containing several generators. It receives a set point and unfairly distributes it by its generating units, observing operating and maintenance restrictions of each unit. This auxiliary function is called load distribution. (Apostolopoulou, 2014).

Operator Power Flow - OPF

The operator power flow, also called optimal power flow, is a tool available in EMS systems commonly used at the level of electrical planning but can be used by the operation as part of network analysis programs. It tries to find the steady-state operating state and optimizes an objective function seeking simultaneously to meet a set of physical and operational constraints established by the operator. These constraints are based on the electrical system under observation behavior knowledge. It considers operational and maintenance restrictions on the use of the equipment and/or transmission lines at a given time.

A detailed mathematical description by Professor Djalma Falcão on optimal power flow can be found in:

http://www.nacad.ufrj.br/~falcao/coe751/fpo01.pdf

Summary of EMS functions

The figure below summarizes how interdependencies between EMS functions (Ankaliki, 2011). Not all of them run in just one control center. For example, the AGC Module is usually divided between the AGC to establish generation set points at each power plant. This is carried out in a system control center. Set-point value, a percent value common to all power plants, is converted to a base power at each power plant and divided unequally between each group and generator. This last function is though executed at each power plant.

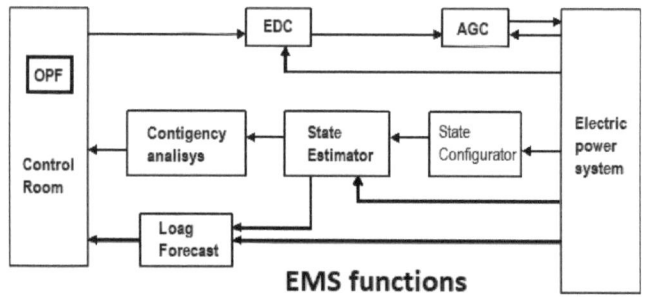

EDC - Economic dispatch calculation
AGC - Automatic generator control
OPF - Optimal power flow

In Brazil, Eletrobras´ Electric Energy Research Center - Cepel offers a range of applications that run off-line under MS Windows environment which are the main functions of an EMS system for use in the planning areas of associated companies such as:

ANAREDE – For the power flow calculation in from a textual or graphic description of an electric system

ANATEN – For the voltage stability analysis

ANAFAS – For simultaneous faults analysis

FLUPOT – For Optimal Load analysis (Centro de Pesquisas em Energia Elétrica - Cepel).

Control centers interaction to other areas

Vast information gathered by the supervisory and control systems was restricted exclusively to the environment of the operating room control rooms. There have been security allegations to disclose a company operational habit.

The possibility of cyber invasion or power systems sabotage was gradually disappearing to allow for information sharing with specialists of other areas or even other companies. Until nowadays, nevertheless, we find electrical power system information restricted only to operation centers control room. These fears, however, were not totally unfounded (Greeberg, 2017).

Reporting assistance

Historical information collected by the supervisory systems and available at the control centers could be spread as operational reports to all company personnel. They should feed the company's management and provide information to electrical planning personnel that works at the company headquarters and rarely in the area or regional control centers. Even non-operator personnel crowded into regional centers usually do not have full access to control rooms.

Service Orders – S.O. preparation assistance

The analysis of logical historical data generated by supervisory systems is used for preparing maintenance service orders and their execution schedule in the field. For example, a field bouncing relay generates an enormous number of text lines inside the SOE - sequential of events, or data - logging files. They should be filtered by computer means before inserting them into a report. They should be filtered as soon as possible. Therefore these reports should be consulted periodically by the maintenance personnel responsible. Periodic reports containing lists of invalid analog variables should regularly feed maintenance and instrumentation experts into the field to increase the security of information generated by control centers.

Most SCADAs do not offer these data warehouse functions and automatically send them to specialized instrumentation personnel. Thus, it is the responsibility of the personnel in charge of the computer maintenance of the SCADA to promote the automation of processes for collecting and sending information to maintenance and instrumentation specialized personnel.

Instrument maintenance personnel are not directly concerned with electrical network aspects but with their components, meters, instruments, and equipment. It is our point of view that SOE should not serve as ammunition immediately for such personnel. The behavior of measuring instruments and equipment statistics is appropriate for such personnel. It is much more important to know how many times a breaker has maneuvered in a given period than at what dates this occurred. It is more important to learn that a relay is changing its state untimely in the field than to know its function. We have not yet had a chance to see SOE data warehousing systematically done in a control center.

Connecting supervisory network to INTRANET and INTERNET

Connecting an installation with lots of equipment to the local control center makes a local network, LAN. The integration of the various local control centers to the area control centers or to the regional control centers and their integration to the system control center make a wide-area-network, WAN. Power companies are known for their resistance to sharing information. They are still known for their prudence in integrating their supervisory systems with other systems to prevent undue intrusion. Thus computer networks that constitute control centers are often isolated from the administrative network available through a company LAN and much more to the World-Wide-Web - WWW network.

Enterprise information sharing

The interconnection of networks of the various energy concessionaires in a national network, known as the National Integrated System, allowed reducing investments in the production of electric energy. This interconnection, however, required companies to transfer information beyond the interconnection points to improve the reliability of the EMS environment to each other. This means that information has begun to spread. Currently, in Brazil, companies communicate from the IT point of view only to the National Operator, and this one transfers needed variable values to each of them.

Offering services on the INTERNET

Pressure has been increasing to offer online and real-time critical information to specialists in the operation and protection of electrical systems not necessarily present inside the company. Thus some companies started creating one or another point of contact of their LAN or WAN network to World Wide Web network. Real-time single-line diagrams, access to company data-logging, real-time warnings of occurrences via SMS and e-mail are already being deployed by some companies in Brazil. An example can be found in (ELETRONORTE, 2018). Naturally, access to services may be only accessed by authorized staff, but a small sample of what can be offered can already be seen on the above link.

Integration to a phasor measuring network

Currents, voltages, powers measured values are approximately sinusoidal waves, $A.\cos(\omega t + \Theta)$. As A, t, and Θ are variables, the representation of an electric quantity of an electric power system sine wave is represented as a complex number and as a vector amplitude and phase-angle, as illustrated in the figure below.

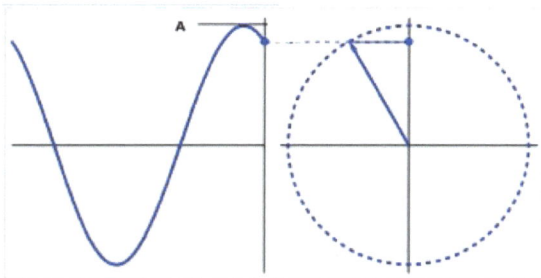

A sinusoidal and phasorial variable representation

An electric quantity can be represented as a complex number by its real and imaginary amplitude and phase angle. For example, in electric power systems, we represent the power flowing in a line by its active and reactive components $P + Qj$, where P is the active power in MW and Q is the reactive power in MVAR. It is usual that bus bar voltages be represented by their module and phase angle.

Phase meter units, PMUs are instruments available for energy companies since the end of the last century. They are multi-meters. They measure several quantities at the same time. Measurements theoretically occur simultaneously and periodically, and the measurement instant is associated with each measured value. The measurement instant is obtained through IRIG-B communication or other protocols. Thus several geographically separated PMUs are GPS synchronized. They are modern instruments, and the measured values are already much better than those obtained by the old traditional measuring instruments. Sampling period can be selected ranging from 10 samples up to 60 samples per second. Some PMUs can achieve even smaller sampling intervals. Sine information is then labeled with sampling date, modulus, and the phase angle relative to a fixed time base and buffered ready to be sent.

The 2005 IEEE C37.118 RFC set the standard for the construction and functionality of phasor measurement units or PMUs.

If we remember that one of the objectives of the state estimator is precisely the calculation of the module of the voltages in the bars and their respective phase-angles, energy companies began to covet the use of PMUs in their facilities and simplify the state estimator algorithms. Of course, since supervisory systems work with much less sampling rate given the amount of information they process interest would be only to use one in every 10 to 60 PMU samples per second.

Experiments used by the author showed that the values obtained from PMUs were much closer to the estimated values than those offered by the old SCADA measurement system. Thus it is only natural that part of the traditional measurement of today's SCADAs comes from PMUs. Some of the PMUs most used by the Brazilian Electric Sector are manufactured by Reason, a company of the General Electric group. It is natural then to expect that the PMUs' measurements be integrated into the control centers.

A new type of assistance center in the operation of electrical systems has been emerging since the availability of the first PMUs. They are the Synchronous Phasorial Measurement Centers, SPMC. Here, a computer or a cluster, that is, several computers interconnected by Gigabit networks behaving like a single computer are being used now, connected via TCP/IP with PMUs scattered throughout electrical installations. Inside these clusters, specialized software packages are in charge of reception, sorting, storage, and phasor data diffusion from PMUs.

They make up the core of specialized operating centers for automatic electrical systems restoration. It uses various strategies, including built-in artificial intelligence, which complements the Synchronous Phasor Measurement Center - SPMC. SPMC begins to appear in the electric system scenario. The traditional operation centers, with the EMS (energy management system) environment discussed in this document, are some of the main clients of the SPMC. Usually, it is these SPMC and not the PMUs that interconnect to control centers.

An example of a free SPMC running under Microsoft Windows environment was developed by the "Tennessee Valley Authority," an electricity company serving the Tennessee Valley in the United States of America, the open PDC software. PDC is the acronym for Phase Data Concentrator. It is open-source software available for customization by the client, capable of obtaining, ordering, storing, and diffusing, using UDP, a TCP mode, for communication with PMUs. A complete discussion on the use of PMU as a tool for monitoring electric power systems in large areas can be found in (Monti, 2016). A discussion of the subject can be found in (Marinelli, 2017).

The first attempt to use phasor measurement to aid in the operation of electrical systems was the production of vast statistical data, an invitation to the "Big Data" techniques to act. The angular difference between meters at various points in the electric grid is a strong indication of the electrical stability operating point of the electric system. The increase of the angular differences of the meters with respect to, for example, those located in the generation indicates that the system is going in the direction of electrical instability.

Some animations have already been produced by academic institutions showing clearly that next to blackouts the phase angles begin to launch in different directions frantically. This is an indication that it is time to act to normalize the system. The problem is the term "frantically". We can currently just watch the disaster happening. Nothing or almost nothing of technical development was done until now for a systemic protection action that avoided a single blackout. (Greene, 2013) & (Monti, 2016)

Operators training

In system control centers we usually find an adequate environment for operator's training. Separate spaces for classrooms and instructors rooms are used to create a suitable simulation environment for operator's training. Computers used for training have real-time access to the electrical system so, that the training environment is more realistic. Electrical system instructors with experience prepare processes scenarios and situations in real-time, generating contingencies for the training operators to evaluate and act upon. The computational effort to obtain these functions is reasonably large. OTS is nowadays an efficient tool for the operator's training.

Integration to business management system, BMS

Optimum load and generation forecasting studies, both mentioned later, indicate the need to use information present in the supervisory systems, and use them in the business planning worksheets and vice versa. Large business management environments like (SAP/R3) used by electric power systems companies must be supplied with data and information originated in the supervisory systems. Enterprise business decision reports, in turn, should be known by the System Operation Centers. The integration of these different but complementary types of entrepreneurial efforts is desirable if not essential today. (SAP/R3, 2018).

Operation control Systems and BMS interaction

The relations between participants of the energy market, that is to say, between the various generating companies, transmitters, and distributors of energy use a BMS environment. The exchanged information is "energy offers," quantities, prices, billing, payments, etc., allowing a business environment to work.

Business Management System (BMS) is the environment where entities such as long-term or short-term contracts, purchase and sale of energy, billing, and business performance evaluation occur. A BMS can be physically integrated into an operating control system, but from the practical point of view, they are always in separate computing environments and interconnect via TCP/IP. Connection is made to various entities, not just one.

To feed these entities a flow of information such as the scheduling of contracts in the BMS => Operating system direction is also essential. They provide the operating system with operational constraints of load forecasting, AGC adjustments, etc.

In the sense Operating system => BMS, elected electric information produced by EMS, such as by the state estimator, contingency analysis and SCADA information, AGC programming and performance and load forecasting reports are delivered to BMS.

Finally, companies transmit and receive data from the various SCADA and EMS functions and administrative computing environments such as (SAP/R3, 2018).

The figure below summarizes the architecture of a typical control center of electrical power systems and its connections with other systems.

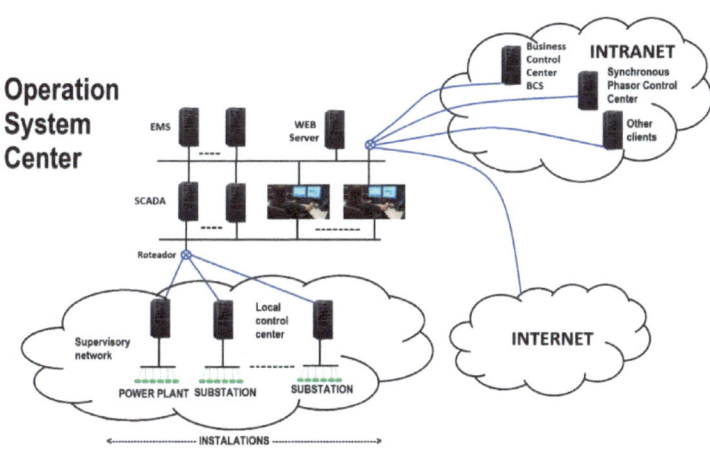

The operation system and its interconnections

The use of modern mathematical tools

Modern mathematical optimization techniques have been proposed for large electrical systems that make use of massive computational performance. Terabytes of data processing have been employed. Thousands of probabilistic simulations hours of operation performed in a few hours involving at least a hundred million major optimization problems have been spent. (Linear mixed integer programming) (PSR, 2018).

Use of artificial intelligence, AI techniques

The most widely used programming language in SCADA is the C language, at least near the core of SCADA. In most traditional control centers, FORTRAN code is still found in part of the EMS programs, especially in the code of power flow calculation that has traditionally been used to make the code. Thus, artificial intelligence techniques have only been used in the last two decades since the beginning of the 21st century. It is only recently that code segments begin to appear written in more modern languages such as python, more suitable for the AI algorithms implementation. Python offers an extensive repertoire of specialized libraries to train neural networks, gradient descending algorithms, support vector machines, etc.

The use of artificial neural networks, ANN

Energy scheduling such as load forecasting is a natural environment for the use of artificial neural networks. Specialized networks in recurrent self-learning of historical series for their training and production of forecasts of short (hours), medium (days), and long term (months).

We found an approach of artificial neural networks implementation in a former operation center embedded in the Canadian Unix-like operating system called QNX. It was an ANN where its learning segment received information from the SCADA or EMS database and, once trained according to desired rules, produced outputs to internal SCADA calculated variables. These resulting variables could be used in reports in single-line diagrams, etc., like any other ones. This provided a generic environment, for example, to produce alarms no more only generated by arriving variables but produced by the ANN application trained for such. The scope of such an ANN application in a control center is unlimited. It was the result of a master thesis work.

In addition to load forecasting, an artificial neural network is an appropriate tool for three-phase fault classification using both supervised and unsupervised ANN. Stability analysis is another example of use with probable success. Kohonen type ANN to address economic dispatch problems and power system stabilizers using recurrent neural networks are cited in (Aggarwall, 1998).

The use of outliers detection techniques

Outliers detection or discrepant or atypical values detection are class of AI that looks for something strange or an object or even a highlighted value outside the main system body or set that is not humanly perceived. In data mining, outlier or atypical detection is a technique for identifying rare items or subsets whose common characteristics do not seem clear when observed in the whole universe. It is in the class of unsupervised learning techniques. There are data to be learned but no examples of output data that fully describe the phenomenon behavior. The careful observation of the results produced by atypical detection techniques can lead to surprising conclusions for phenomena associated with large-dataset.

An example of atypical detection in power distribution control centers is to find consumer fraud by observing the time series of the monthly KWh consumed. Integrated systems in these centers can systematically acquire the history of each residential or industrial consumer and discover periodic anomalous behaviors that may indicate fraud or energy meters malfunctions.

Outlier detection could also be used as a pre-processing or data warehouse process in the data before executing the state estimator, for example, providing for the exclusion of absurd values, thus improving performance.

A frequent example in artificial intelligence courses is the application of outlier detection for emails spam analysis. This technique can be adapted to the treatment of alarm lists in control centers to identify anomalous or unexpected behavior that does not match the global philosophy adopted in the protection system design by the undue performance of protection relays.

A simple application of atypical detection to automatically establish the reasonable and a priori operational limits of the variables in a supervisory system can be found in (Martins, 2016).

Stability assessment using decision trees

A specific example of the use of decision trees was suggested by (Ubiratan H. Bezerra, 2017). Here successive samples of the results of chained load flow aggregated to the values of the states of the main equipment in a supervision system are trained by a decision tree algorithm to produce situations in which the electrical system may be heading to an instability region. One of the results of this proposed methodology is to alert the electric system operator to specific areas approaching risks of the electrical system by focusing its attention on areas of the system that are more likely to compromise the system. The number of required variable values that operators monitor in their daily work may lead them to not perceive regions of the electrical system with the highest operational risk. This proposal addresses such a problem. (Ubiratan H. Bezerra, 2017).

The high number of samples, Big Data

Big Data is the term currently used when referring to such a data set that current technology does not allow its treatment using methodologies, procedures, and techniques in use until then. Data tables are usually treated as in databases. They are stored in organized spaces and managed by specialized programs such as Oracle, PostgreSQL, MySql, Microsoft Dbase, etc. Data is stored in a sophisticated way, known only from the program that treats them. Consultation access and updating are performed by a practically universal language - SQL - Structured Query Language. This language organizes the storage, maintenance, and retrieval of information through relationships between them. Their price, however, is high required storage space and speed of access as the number of information increases.

Thus for a large amount of data, its treatment becomes unfeasible by traditional procedures. The amount of data currently stored by control centers elects them to have their information handled by modern techniques and may be considered candidates for treatment with "big data" techniques. A control center supervising about forty to fifty substations handles more than 100,000 variables. In general, only less than 20 % of the information is analogical information and therefore periodic. The other 80 % is logical information and therefore of random occurrence. Considering that information from the field is stored along with the date of arrival at the control center or the date of its acquisition, A historical database, even using textual storage, can reach hundreds of thousands or millions of rows. In a system operation center, this number reaches the order of a billion easily.

Renewable energy

A brief discussion of renewable energy, also enthusiastically referred to as clean energy, and its relationship to electrical power systems control centers follows. A control center was synonymous with the term "load dispatch" for a long time. This denomination emphasizes one of the biggest concerns of operators or "dispatchers," as they were also known. Load dispatch is to conveniently use generation resources, their availability, and their constraints to meet the need of load demand from consumers. Dispatching, therefore, means making decisions, for example, which generators of a hydroelectric plant should come into operation, which thermal plants can be turned off to save on fuel? The availability aspect here is of fundamental importance. We will briefly discuss each type of renewable energy and its interaction with a control center.

Solar energy

Photovoltaic panels are being installed in large quantities worldwide. A significant contribution of energy production in the world is already in charge of photovoltaic generation. Large farms of photovoltaic panels can be connected to control centers for convenient dispatch of available energy. Unfortunately, we cannot use solar energy in the evening and even during sunset, when peak demand occurs. The use of batteries seems imperative. But its buying cost, maintenance costs, fragility, and durability are still a problem to overcome.

In Germany, the Government has been encouraging for some time the installation of photovoltaic panels on the roofs of homes and buildings of companies and public buildings. These arrangements are authorized and encouraged to interconnect with the utility grid to provide power to the community when not internally used. The use of batteries to maintain in the evening is not necessary. Simply use your local dealer's supply to replace residential photovoltaic during the evening and sell any surplus power during the day.

That is, the distribution company functions as batteries for the domestic solar panel systems. In Brazil, bureaucratic efforts to be overcome equipment costs still discourage initiatives in this area. This energy supply is not dispatchable and therefore is not controllable. This is not yet completely solved. It is eventually "felt" in distribution control centers as negative loads.

Large solar farms are usually integrated with transmission companies' control centers. However, domestic solar generation is a problem linked to control centers of distribution companies. The equitable buying and selling of solar energy are claimed by these companies as unfair. At first consumer/generator only agrees to buy paying the price it sells. The remuneration of the surplus sold is still subject to discussion and regulation in Brazil.

Wind energy

Another clean energy source that has been in use for some time is the energy of the winds. Large wind farms are being set up in high regions onshore and offshore, mainly in the Nordic countries. This form of energy production, however, is dispatchable. Wind farms can be integrated into control centers for both supervision and control. In general, energy converted from the wind is produced in direct current because the speed of the blades of the generators cannot be controlled with precision. Later it is alternated so that it can be synchronized and connected to the power grid.

Unlike hydraulic power, the wind does not blow continuously. Usually, it offers itself in bursts which translates into offers of energy not as constant as that produced by the flow of a river. Thermal power plants are the most available sources of energy. Just keep them fueled, day and night, whether coal or diesel oil. An old electrical systems engineer told me at a congress in Sao Paulo that the absence of wind brings happiness to power system operators.

Wind farms are usually considered as negative charges by control centers. More recently, variable speed turbines offer the possibility of controlling the active and reactive power of the entire farm by adjusting these parameters for each one.

Integrating wind farms and control centers is, therefore, a necessity. However, specific strategies are beginning to be developed about the types of information that must be exchanged. The primary energy source, the wind, does not have a very smooth behavior in the short term. Thus, wind speed is a variable that must be included in the control strategy in energy produced by wind farms. For hydraulic plants, for example, the speed of water is not frequently taken into account.

The question for the dispatcher is whether or not count on the energy source as an available source at any moment or forgets and considers it as a negative charge. The availability of a turbine's power supply is entirely tied to wind speed. The behavioral model commonly used for power supplied versus wind speed is illustrated in the figure below.

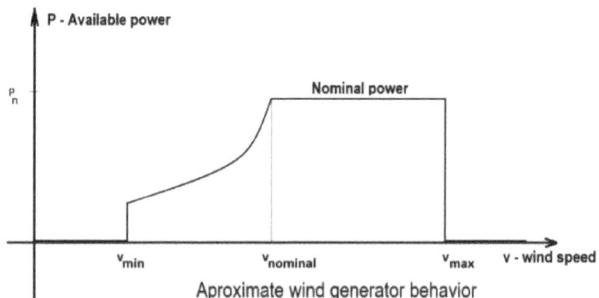

Aproximate wind generator behavior

When wind speed is below the minimum specified by the manufacturer, the strategy is to turn off the turbine to avoid the risk of mechanical damage. Likewise, for winds with speeds higher than the maximum value, the turbine blades are also locked. Between the recommended nominal velocity and the maximum allowed, the turbine offers a constant electric power value. Here it behaves exactly like a hydraulic turbine in mighty rivers. The biggest problem to be faced by the strategy of using wind farms in control centers appears when the winds are blowing with speeds between the minimum value and the nominal value.

Other sources of energy

Biomass is not yet of significant importance from the point of view of operation centers since they are still mainly used in agricultural farms. Few government projects already use urban waste to produce electricity.

Tide or wave energy, having Portugal as one of the leading countries, takes advantage of the fact that geographically, the country is very much in the Atlantic Ocean. Tide or wave energy can also be dispatchable with some difficulties inherent to the intermittency of the waves and the flow of the tides.

The smart grid

Computer software embedded in protective, measuring, and control equipment may not yet be considered "smart." However, modern communication systems, especially optical fiber ones, allowed the intense exchange of messages between the processed equipment and help operational decisions. So, it is apparently intelligent. So, the grid is called "smart" today. That is how the electric grid became a smart grid. (US Governement)

The blackouts on August 14, 2003, which affected some 10 million people in Canada and 45 million people in the United States, similarly to what had happened in 1965, already mentioned, gave birth to EMS, caused a paradigm change again in the conduct of electric power systems. The development of new government recommendations pointed to the need to distribute intelligent computer systems associated with electrical equipment and instruments offering smarter services and using the WEB as a means of communication. The network or grid was born in the 1890s. It is basically a network of interconnected electric wires and cables, a natural path for low frequency and high-frequency electric current, which is also suitable for the transport of information. We would also like to remind you that current networks already offer information transport through their OPGW optical grounding cables, a true backbone for TCP/IP. So, it is natural to embed "intelligence" in your equipment and instruments to improve the reliability, safety, and availability of electricity to consumers. These recommendations gave rise to the smart grid and gave birth to a new area of human knowledge, grid computing.

The fact that the evaluation of that blackout highlighted was that operators of control centers of companies not directly affected were not aware of what was happening. Thus valuable and timely dispatch arrangements that could have reduced the impact of the oscillations did not occur. The recommendations then pointed to the urgent need for national information exchange among all control centers so that all operators can access the general framework or the state of the interconnected electrical system in the country.

The smart grid is not one piece of equipment, or a computer program, or a computer system. It is a set of all that has been implanted gradually, instrument to instrument, installation to installation, control center to control center, and in each of the companies of production, transmission, and distribution of electric energy. The smart grid is a set of goals that have been pursued ever since.

As far as consumers are concerned, replacing their old residential and industrial meters that only reported a KWh counted by "smart" energy meters will allow efficient consumption control. For example, a power meter can send an SMS informing you that your consumption is too high for the hour of the day, or it can even decide to ask the refrigerator to switch off for a few tens of minutes if its internal temperature allows.

Why not concede car batteries to be charged at dawn during off-peak hours, as historically the owner usually does not leave the car early in the evening during Mondays to Thursdays? These are symbolic examples but current technology in dedicated microprocessors such as Arduino, for example, already allow this technology deployment these days. Modern homes and industries already offer built-in Wi-Fi networks, thus promoting the communication infrastructure required for "smart" equipment to exchange information. Of course, this structure also provides hackers promising scenarios.

Concerning control centers, for example, the intelligent exchange of information between protective equipment can lead to decision-making dispatch, or possibly, the islanding of system areas by consulting the energy availabilities and load priorities during a located blackout.

Phasor measurement centers already mentioned earlier are examples of contributions to the smart grid. The objective pursued here is quite daring and complex. It aims to achieve a network that is automatically restored without human participation, quickly and safely, if possible.

The considerable increase in alternative sources of energy such as wind and photovoltaic, as discussed previously, has increased both the amount of electricity generation facilities that an efficient and safe energy dispatch is practically beginning to be humanly impossible. The sum of all these technologies and efforts has been building the smart grid.

However, up to these writings, the computational effort on the smart grid has not yet achieved broad and effective control objectives. We continue to be disaster observers. Old, traditional control centers still do not have this responsibility. They are independent control centers. We expect that synchronous phasor measuring centers do it. (PSR, 2018).

A future Electric Power System Control Center

Final considerations

We hope that this book, already obsolete in its birth, be fruitful to electrical engineering students, electrical technicians and data processing staff, and other practices, interested in working with power systems. Whether at fieldwork, not at all advisable or in control centers or in industrial automation in general, supervisory and control systems, whether electrical power systems, manufacturing industrial environments, petroleum, or chemical in general, require multidisciplinary knowledge.

The operation centers crew must understand the common jargon, and some rules specific to the various areas of knowledge involved, for a good human relationship. We hope that this book can help reduce the height of this Babel Tower.

Digital technological evolution in its exponential trajectory was the decisive factor for the technical advance in the Electric Sector as in many other areas of human activities. It was the decisive factor to increase the worker's quality of life in the Electricity Sector, especially for the general population, reducing the risk of material and especially human losses, besides improving efficiency, reliability, and safety in the conduction and electric power systems operation is the goal.

No matter the technology is available to make life easier and improve business and human productivity. No matter the efforts to produce massive electricity, without which, we would have hard-living, good people relationships, and understanding, resilience is, in this author's point of view, understood as the major factor for human evolution and well-being. A current description of the subject is found in (Government).

BIBLIOGRAPHY

Power System Load Flow Tutorial Part 1. (2018). Source: https://www.youtube.com/watch?v=LeGss3hdpMs

Protocols applied for time synchronization in a digital substation automation. . (2018). Source: Electric Engineering Portal.: https://electrical-engineering-portal.com/time-synchronization-substation-automation#irig-b

Aggarwall, R. &. (1998). *Artificial neural networks in power systems. III. Examples of applications in power systems.* Source: https://ieeexplore.ieee.org/document/745267

Ankaliki, S. (2011). *Energy Control Center Functions for Power System.* Source: https://www.academia.edu/32197974/Energy_Control_Center_Functions_for_Power_System

Apostolopoulou, D. (2014). *Automatic Generation Control and its Implementation in Real Time.* Source: IEEE Computer Society: https://experts.illinois.edu/en/publications/automatic-generation-control-and-its-implementation-in-real-time

Barros, T. M. (2014). *Previsão de carga – Comparação de técnicas.* Source: Faculdade de Engenharia da Universidade do Porto.: file:///C:/Users/odani/AppData/Local/Temp/143389111.pdf

CEMIG. (2009). *Indicadores de Qualidade.* Source: https://www.google.com/url?sa=t&rct=j&q=&esrc=s&source=web&cd=&ved=2ahUKEwi9veDP_OLyAhU9K7kGHbhfDCAQFnoECAMQAQ&url=https://www.cgti.org.br/publicacoes/wp-content/uploads/2016/01/Monitoramento-da-Qualidade-da-Tensão-através-do-indicador-CQT.pdf&usg=AOvVa

Cepel. (2018). *Análise de Redes Elétricas.* Source: http://www.cepel.br/pt_br/produtos/programas-computacionais-por-categoria/analise-de-redes-eletricas.htm

Corsi. (2015). *Voltage Control and Protection in Eletrical Power Systems.* Source: https://www.amazon.com.br/Voltage-Control-Protection-Electrical-Systems-ebook/dp/B0101JVRU6

ELETRONORTE. (2018). *Sistema Elétrico Eletrobras Eletronorte.* Source: https://eletro.eletronorte.gov.br/

Greeberg, A. (2017). *How an Entire Netion Became Russias Test Lab for Cyberwar.* Source: https://www.wired.com/story/russian-hackers-attack-ukraine/

Greene, B. K. (2013). *Novel Applications for Phasor Measurement Units and Synchrophasor Data.* Source: https://www.semanticscholar.org/paper/Novel-Applications-for-Phasor-Measurement-Units-and-Greene/0c67ad6596124d55f168935264e656d3a85807d2

Marinelli, M. (2017). *Demonstration of visualization techniques for the control room.* Source: http://orbit.dtu.dk/files/130245204/D8_1_Demonstration_visualization_for_control_room_2030.pdf

Martins, D. A. (2016). *Resultados de Análise para Deteção de Outliers Aplicados a Dados de um Sistema Elétrico de Potência.* Source: https://marajo.xyz/OUTLIERS/

Microsoft. (2016). *Coding Guidelines.* Source: https://docs.microsoft.com/en-us/cognitive-toolkit/coding-guidelines

Mishra, V. &. (2012). *Contingency Analysis of Power System.* Source: http://citeseerx.ist.psu.edu/viewdoc/download?doi=10.1.1.736.3319&rep=rep1&type=pdf

Monti, A. M. (2016). *Phasor Measurements Units and Wide Area Monitoring Systems.* Source: Elsevier: https://scholar.google.com.br/scholar?q=Phasor+Measurements+Units+and+Wide+Area+Monitoring+Systems.+Elsevier&hl=en&as_sdt=0&as_vis=1&oi=scholart

Monticelli, A. (1999). *State Estimation in Electric Power Systems.* Source: Springer: https://www.springer.com/gp/book/9780792385196

Monticelli, a. (2011). *Introdução a Sistemas de Energia Elétrica.* Source: https://www.springer.com/gp/book/9780792385196

PSR, C. -T. (2018). *Novo modelo de planejamento energético da Costa Oeste dos Estados Unidos.* Source: http://www.psr-inc.com/noticias/?current=p11463

SAP/R3. (2018). *SAP Business One Cloud - Para empresas em crescimento*. Source: Economic Dispatch: https://www.sap.com/brazil/cmp/dg/br-sap-business-one-brasil/index.html?campaigncode=CRM-BR21-PPC-GBUBUSB&gclid=CjwKCAjwj8eJBhA5EiwAg3z0myVX0HJ9ijGv4G9EQr8BXqIdM0YKf87RXarOCimHzW60eoebxdqz4R0CHUQQAvD_BwE&gclsrc=aw.ds

Ubiratan H. Bezerra, J. P. (2017). *Metodologia de Controle Preventivo Baseada em Árvore de Decisão para a Melhoria da Segurança Estática e* . Source: IX CITENEL Brasil: https://www.researchgate.net/publication/319087816_Metodologia_de_Controle_Preventivo_Baseada_em_Arvore_de_Decisao_para_a_Melhoria_da_Seguranca_Estatica_e_Dinamica_do_Sistema_Interligado_da_Eletronorte

US Governement. (s.d.). Source: SMARTGRID: https://www.smartgrid.gov/

Wikipedia. (2018). *Economic Dispatch*. Source: Wikipedia: https://en.wikipedia.org/wiki/Talk%3AEconomic_dispatch

Wikipedia. (2018). *Network Protocol*. Source: https://en.wikipedia.org/wiki/Internet_Protocol

Wu, F. K. (2005). *Power System Control Centers: Past, Present and Future.*. Source: Proceedings of the IEEE: https://ieeexplore.ieee.org/document/1519722

About the author

Daniel Augusto Martins is an electrical engineer, electronic option, graduated from the Federal University of Pará in 1972. He holds a master's degree in control systems sciences from the Federal University of Santa Catarina in 1974. He was a teaching assistant at UFSC and an assistant professor at UFPA.

He worked in the North-South Electric Power Plants, Eletronorte from 1979 to 2012 where he engaged the teams that developed the software for Tucuruí Hydroelectric Power Plant operation center at CGEE Alsthom in Paris.

He was a leader in the development and implementation of the Eletronorte´s System Operating Center in Brasilia and participated in the implementation of the Regional Operation Center of Belém.

He is a programmer specialized in software development for industrial automation and a Jurassic programmer in ANSI-C, Assembler, FORTRAN, BASIC, PHP, JavaScript, etc. Daniel is an enthusiast in the development of WEB pages with dynamic interface to industrial processes.

Daniel is retired and lives in Belem, north of Brazil, with his Maria, three children and four grandchildren. More information about the author can be found at https://marajo.xyz/daniel.php.

www.ingramcontent.com/pod-product-compliance
Lightning Source LLC
Chambersburg PA
CBHW040315220526
45473CB00009B/2444